MATHEMATICS EDUCATION

Mathematics Education offers both undergraduates and starting-graduate students in education an introduction to the connections that exist between mathematics and a critical orientation to education. This primer shows how concepts like race, class, gender, and language have real effects in the mathematics classroom, and prepares current and future mathematics teachers with a more critical math education that increases accessibility for all students. By refocusing math learning towards the goals of democracy and social and environmental crises, the book also introduces readers to broader contemporary school policy and reform debates and struggles.

Mark Wolfmeyer shows future and current teachers how critical mathematics education can be put into practice with concrete strategies and examples in both formal and informal educational settings. With opportunities for readers to engage in deeper discussion through suggested activities, *Mathematics Education's* pedagogical features include:

- Study Questions for Teachers and Students
- Text Boxes with Examples of Critical Education in Practice
- Glossary

Mark Wolfmeyer is Assistant Professor in the College of Education at Kutztown University of Pennsylvania.

Critical Introductions in Education Series

Series Editor: Kenneth J. Saltman

The Politics of Education: A Critical Introduction, second edition
By Kenneth J. Saltman

Mathematics Education: A Critical Introduction
By Mark Wolfmeyer

MATHEMATICS EDUCATION

A Critical Introduction

Mark Wolfmeyer

NEW YORK AND LONDON

First published 2017
by Routledge
711 Third Avenue, New York, NY 10017

and by Routledge
2 Park Square, Milton Park, Abingdon, Oxon, OX14 4RN

Routledge is an imprint of the Taylor & Francis Group, an informa business

Library of Congress Cataloging in Publication Data
Names: Wolfmeyer, Mark.
Title: Mathematics education : a critical introduction / by Mark Wolfmeyer.
Description: New York : Routledge, 2017. | Includes index.
Identifiers: LCCN 2016035095 | ISBN 9781138243279 (hardback) | ISBN 9781138243286 (pbk.)
Subjects: LCSH: Mathematics—Study and teaching—United States. | Education—Social aspects—United States.
Classification: LCC QA13 .W654 2017 | DDC 510.71/073—dc23
LC record available at https://lccn.loc.gov/2016035095

ISBN: 978-1-138-24327-9 (hbk)
ISBN: 978-1-138-24328-6 (pbk)
ISBN: 978-1-315-26952-8 (ebk)

Typeset in Bembo
by Apex CoVantage, LLC

For my parents, Helen and Paul

CONTENTS

FIGURES AND TABLES

Figures

Tables

SERIES PREFACE

Mathematics Education: A Critical Introduction is an exciting addition to the Critical Introductions series. Books in the series provide critical introductions to social studies education, math education, English education, science education, art education, educational leadership, and more. The series is designed to offer students who are new to these subjects in education an introduction and overview—a first book for a first course. These "primers," covering the key subjects of education, are intended to help students broadly comprehend their new field socially and politically. While primers in the series engage with dominant liberal and conservative views on subjects, they ask readers to comprehend dominant perspectives of a subject area through a critical lens that focuses on social justice, power, politics, ethics, and history. Additionally, these Critical Introductions provide students with a new vocabulary and key framing concepts with which to interpret future knowledge about the field gleaned through academic study and clinical experiences in schools. For this reason, Critical Introductions include boldfaced key terms in the text that are defined in a glossary in the back. They also include lists of suggested readings and potential questions for discussion accompanying each chapter. The books are suited for instructors to pair chapters with selections from the lists of suggested readings at the end of each chapter. Ideally, these Critical Introductions can be both a kind of field guide or handbook arming students to interpret experiences in schools and serve as a foundational text for future deeper scholarly study and development of a critical understanding of educational subjects built through further engagement with newly acquainted authors and texts. The Critical Introductions series also offers a basis for social and political engagement and activism within the field of education because they ground their examinations of particular subjects in terms of broader contemporary school policy and reform debates and struggles. In this sense, even advanced graduate students and seasoned scholars in education would benefit from consulting these Critical Introductions for a fuller comprehension of their field and the political struggles and stakes in several subjects affecting teachers, students, and colleagues.

This book, *Mathematics Education: A Critical Introduction*, provides an overview to its subject. Mathematics education, perhaps more than any other area in teacher education, is framed as neutral and apolitical and has been codified in ways that resist examination of the values, assumptions, interests, and ideologies that organize knowledge and pedagogical practice. Mark Wolfmeyer breaks through the assumptions that math education is neutral, disinterested, and universal and instead gives the reader an understanding of how math education relates to social and racial inequality, gender disparity, and class oppression. His account clearly and accessibly investigates the philosophical underpinnings behind math education in its traditional and critical variations. As well, in the course of showing what is entailed in a specifically critical math education, he engages with the different perspectives of major scholars. His book explores the broader economic, social, political, and cultural implications of teaching math and makes these concerns central rather than incidental to mathematics education.

Wolfmeyer's work is unique in helping new math teachers ask the question of why they are becoming math teachers, what it means socially, and how it affirms or contests existing social arrangements. Teachers, through their practices, inevitably make meanings in the classroom, and the meanings that they make have social import. Wolfmeyer shows that it is impossible for a math teacher to be outside of politics in this sense—that is, math education is implicated in the inevitable conflicts among groups and classes over money and resources but also over symbols, values, and meanings. The crucial matter is whether or not one comes to understand these inevitably political dimensions of teaching math. Moreover, once one comes to understand that teaching math is inevitably political, Wolfmeyer shows the reader what s/he can do about it by making the teaching of math a force for challenging oppression and exploitation. Wolfmeyer offers a vision for math education as a force for expanding justice, equality, fairness, and inclusion.

PREFACE

This book presents my attempt to bring together the most critical work in mathematics education and make this accessible for future and current mathematics teachers. I also hope that critical educators more generally will find it helpful in understanding mathematics teaching. What, you ask, do I mean by critical? I start my answer with two famed scholars of mathematics education, Ole Skovsmose and Brian Greer (2012), who first argue that to be critical is to challenge. This is the conventional use of the term; to be critical is to ask questions, to check for hidden assumptions, to push and prod. But second, in looking at etymological relations, they also remind us that to be critical is to *attend to crisis*. True to this dyad, then, the assemblage contained herein opens up mathematics education for its contribution to the crises of our time, as well as the opportunities existing within mathematics education that can interrupt them.

Take the following as examples of modern-day crises: racial injustice, gender inequality, social class hierarchy, and environmental catastrophe. As will be revealed in the contents of this book, to think critically about mathematics teaching is to examine the underlying sociopolitical orderings of relations between groups of people through a focus on power, ethics, and historical and cultural understandings. Advanced work in critical social theory suggests that such a framework illuminates the aforementioned crises of our time. Sadly, mathematics education as it is largely practiced reinforces these unjust circumstances and, interestingly, does so with a veil of neutrality. The propagation of mathematics as an objective, value-free discipline will be our first line in critiquing mathematics education as it is typically conceived. In place of this, we can view mathematics as a socially developed collection of not-yet-disproven concepts, and such a view begins to open our eyes to the manner in which a mathematical education can interrupt today's crises.

After an introduction to the philosophy and anthropology of mathematics, the second chapter introduces the first step toward teaching mathematics critically, that of reform mathematics education. This orientation will remain present throughout the book as we discuss such a pedagogy's promises and limitations. In the third, fourth, and fifth chapters, we take the social constructs of race, class, and gender, respectively, in turn. Each of these three chapters moves through the relevant critical social theory before engaging with advancements in mathematics education on the topic. In all, there exists the dual objective of critiquing mainstream mathematics education as well as redefining it for critical work.

Regarding these efforts, I feel compelled to provide some words of caution. First, treating each topic (race, class, and gender) on its own presents a certain danger, namely that singular discussions focusing on one social identity at a time might cause us to have a narrowed, incomplete picture or perhaps privilege one factor (say, social class) over another. For this reason, the concluding chapter makes important the notion of intersectionality, an advancement in social theory in which the interrelated natures of race, social class, gender, and so on are highlighted. Another limitation of the discussions here is the imbalance in space devoted to the differing crises. I chose to write entire chapters devoted to race, class, and gender mostly because these have been attended to significantly by critical educators of mathematics; unfortunately, much more work is to be done on disability studies, language-minority students, and sexuality and mathematics education, for example. I do, however, touch on these topics as well as the environmental crisis where I found it relevant and hope that future efforts in critical mathematics will attend to these issues.

I close with my intended audience for this book and an introduction to its features. First and foremost, I wrote it with future mathematics teachers in mind, and I plan to use it for undergraduate and early graduate students. In teacher education, I suggest that it be used as a text in either a mathematics pedagogy course or an educational foundations course. The book's style is conversational, and it contains features to aid in accessibility, including the glossary at the end of the book, and at the conclusion of each chapter you will find suggested activities and prompts for discussion. In these efforts, I also expect that current mathematics teachers eager to deeply examine their practice will find the book easy to use. Finally, to broaden the readership, I also made sure that mathematical discussions do not require advanced prior knowledge of mathematics. This provides greater access to critically teaching mathematics for other readers, such as students and scholars of educational foundations, who might come to the conversation with a critical rather than mathematical orientation.

Reference

Skovsmose, O. & Greer, B. (Eds.) (2012). *Opening the cage: Critique and politics of mathematics education*. Boston, MA: Sense.

ACKNOWLEDGMENTS

First and foremost, I thank members of my family for their support in writing this book. My partner Ellie Escher and children Beatrice and Guy WolfmeyerEscher have provided much encouragement and the space that I needed to get the project done. Thanks specifically to Beatrice for help with Figure 1.3! Also, thanks to my parents Helen and Paul Wolfmeyer, to whom I have dedicated this book. I have the critical orientations needed to write it thanks to their efforts in teaching me justice and empathy since I was a child. Last, to my in-laws Gus and Connie Escher and to siblings David Wolfmeyer, Beth Cocuzza, and Amy Escher, thank you for your support.

Thank you to Kenneth Saltman for inviting me to submit a proposal for the series and for feedback throughout the writing process and to Catherine Bernard, editor at Routledge, for help in shaping the book as it developed and for seeing it through to print. Also continued thanks to my early mentor, Joel Spring, for teaching me a style of writing that increases access to challenging, complicated, and critical topics.

A number of others provided encouragement, support, and/or extensive feedback on this book. These include John Lupinacci, Nataly Chesky, Brian Greer, Erika Bullock, Theresa Stahler, Patricia Walsh Coates, George Sirrakos, E. Wayne Ross, Greg Bourassa, Graham Slater, Miriam Tager, Edwin Mayorga, and Charles Nace. Finally, thanks to the blind reviewers who provided substantive feedback to augment the book's contents and accessibility.

1

WHAT IS MATHEMATICS?

From Mathematicians to Philosophers and Anthropologists

This book provides an introduction to teaching mathematics with a critical perspective. A natural starting place is to looking critically at the discipline of mathematics, and you might be thinking, how can we look critically at mathematics? Do we not consider mathematics to be the objective, value-free knowledge that is free from argumentation and contention? Is it not true that $1 + 1 = 2$ and there is no evidence to the contrary? It turns out that there is much dispute as to the nature of mathematics and what counts as math. Mathematics means one thing to one person and a completely different thing to another.

We answer this chapter's question first with a review of what people typically think mathematics is and next the activity of mathematicians, giving meaning to its two main branches: pure and applied. Next, I introduce the world of philosophy of mathematics for an answer, and this brings forth an interesting question: Does mathematics exist external to people, or did we invent it? Reviewing these philosophies begins a deeper critique of the assumptions we typically hold about mathematics; this will continue with the subsequent introductions to history of mathematics and the burgeoning field of ethnomathematics. Both push us to think of mathematics less as a static world of academic (mostly white and western) development and rather as a multicultural and social activity.

For this chapter, as well as for the whole book, I intend to provide a meaningful experience for readers with a variety of backgrounds in mathematics and education. The discussion on mathematics contained herein is appropriate for those with little or negative experience in mathematics; I also expect readers with a stronger mathematical background to find the discussion fun and instructive. For example, these readers might find interesting the particular choices I made in discussing mathematics and the discussions regarding philosophies of mathematics as well as ethnomathematics.

An Introduction to Mathematical Behaviors and Pure and Applied Mathematics

Let's begin to answer what mathematics is by describing a few mathematical behaviors. First, there is the mathematical behavior of computations with numbers. This is a common point of reference for an understanding of mathematics. You or someone you know of, for example, may say, "Ugh, I am so bad at math!" when confronted with a task that requires multiplication, division, addition, or subtraction of two numbers. The "basic four computations" are likely the first things that come to mind when most people think of math.

Let's play in this conception of math a bit. See if you can add the numbers 781 and 312 without a calculator. Many of you might like to reach for a pencil and paper and set up the problem as you were taught in school. These pencil-and-paper methods are referred to as *standard algorithms* for mathematical computations. At some point, a person (like a teacher or a parent) might have encouraged you to try to answer such problems without the use of the standard algorithm, instead prompting you to develop a reasoned computation strategy. Try the problem again. The goal is not to say the answer (1,093) but to argue how you arrived at the answer. Now try to give a mental computation for 43 times 15. Give yourself a chance to come up with some methods before reading the next paragraph.

There are many ways to compute the answer of 645, and here is one of these as an example. Break the 43 into 40 and 3. You know you need to multiply 40 times 15 and 3 times 15. The second one is easy by repeated addition (45). The first problem can be made simpler again by multiplying 4 by 15 (60) and adding an extra 0, because the problem is really 40 by 15 (600). Now you add your two parts together to get 645. You may be able to follow this short narrative, and/or Figure 1.1 may help. This method relies on the fact that we understand the concept of multiplication. We can think about multiplication as repeated addition. For example, 5 times 4 is 5 + 5 + 5 + 5. This was an important part of our method. For one, we easily saw how 45 is the product of 3 and 15. It also allowed us to break apart the 43 into 40 and 3 and then add the results together.

I have been describing the knee-jerk response to the main question of this chapter: Mathematics means "doing these types of computations." Mathematics, however, is much more than this, and we need to look at the other behaviors that can be considered mathematical. Where to go next but with those whose professional activity centers on such behaviors: mathematicians. To the point, I once took a course called Abstract Algebra with a mathematician who often proclaimed that he was "no good with numbers." He was partly joking, but it is true that he rarely encounters a number in his own research. Right from the start, then, we begin to see how mathematics is far more than the number and operations that quickly come to mind.

Many university math departments, where mathematicians often work, are split into two divisions: pure and applied. In *What Is Mathematics, Really?* Ruben

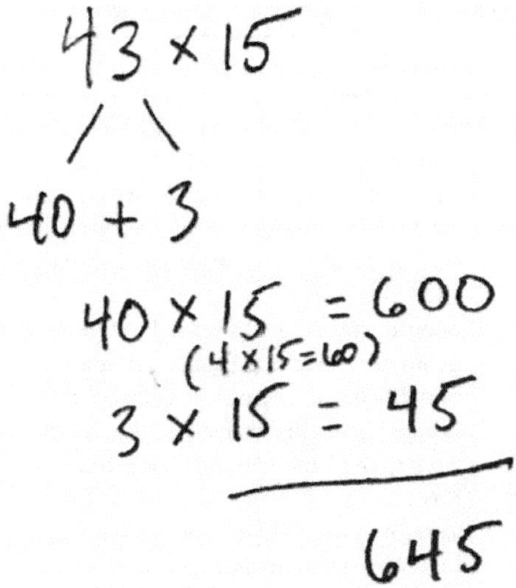

43 × 15
/ \
40 + 3
40 × 15 = 600
(4 × 15 = 60)
3 × 15 = 45
645

FIGURE 1.1 A worked-out mental math example

Hersh (1997) differentiates the two as follows: "Mathematics that stresses results above proof is sometimes called 'applied mathematics.' Mathematics that stresses proof above results is sometimes called 'pure mathematics'" (p. 6). Pure mathematicians work within abstract worlds to prove things that have little association to a particular physical (or social) situation. Applied mathematicians do the opposite: they start with these physical or social situations and adapt the work of pure mathematicians to address particulars within the real-world application.

Within each division are a host of topics. From among the topics in pure math are number theory, algebra, geometry, topology, calculus, analysis, and combinatorics. Table 1.1 gives an elementary description of each of these.

Generally speaking, applied mathematics includes any kind of mathematical knowledge that has made a connection to a real-world problem. Such endeavors have spawned particular fields of their own, such as differential equations, mathematical modeling, statistics, mathematical physics, and game theory. Table 1.2 gives an elementary description of each of these. Both the applied mathematics table and the preceding pure mathematics table give a sense of what these branches are about. In truth, most are difficult to define narrowly, and none are entirely isolated from any other branch. These are partial lists aiming to distinguish the characteristic differences between pure and applied mathematics.

Thus a variety of mathematical topics are at play among the work of mathematicians. The distinction between pure and applied mathematics proves highly relevant as we look at mathematics critically in order to conceptualize how we will teach it and for what purpose. I want to illuminate this distinction with two

TABLE 1.1 Some branches of pure mathematics, with description

Mathematical branch	*Description*
Number theory	Discrete mathematics, the study of integers, primes, rational numbers
Algebra	Study of mathematical symbols and operations; ranges from elementary equations (as in school algebra) to advanced topics, such as linear (vector spaces) and abstract algebra (groups, rings, and fields)
Geometry	Properties and theorems related to figures; spans Euclidean geometry (including elementary topics taught in schools) to non-Euclidean geometries (advanced)
Topology	Advanced geometry that studies figures that are fluid, those that are stretched and bent but not torn or glued, with a focus on set theory
Calculus	Focuses on change with respect to functions, with two main branches: differential (instantaneous rates of change, slopes of tangent lines to curves) and integral (areas under curves)
Analysis	The broader branch of mathematics that includes calculus; focuses on continuous functions and real numbers
Combinatorics	Discrete mathematics; focuses on countability; includes probability and some work in algebra and geometry

TABLE 1.2 Some branches of applied mathematics, with description

Branch of applied mathematics	*Description*
Differential equations	Application of calculus (derivatives) to physics, chemistry, biology, other hard sciences, engineering, economics
Mathematical modeling	Description of a physical or social system using mathematics; useful in studying components of a system and making predictions
Statistics	Collection, analysis, interpretation of numerical data
Mathematical physics	Branch of applied mathematics dealing with physical problems; a type of modeling
Game theory	Study of decision making with applications to many fields of study including biology, economics, and political science

mathematical examples from basic and intermediate mathematics. First, returning to our computation problem, we can see how 43 times 15 is an abstract concept, and our method of computing it required a conceptual understanding that multiplication is repeated addition. It is not too difficult to imagine a context in which we would have to apply such knowledge. For example, I might want to determine how much compost I need to spread on my garden that has dimensions 43 feet by 15 feet. In a sense, multiplication is both pure and applied mathematics.

For another example, you may have learned a bit of trigonometry in your mathematical experiences. One fact in basic trigonometry is about the relationships between the side lengths of a right isosceles triangle, that is, a triangle with one 90-degree angle and two 45-degree angles. If you know one of the lengths of the "legs," the two sides that are equal, then you can approximate the longest side by multiplying the leg length by about 1.4 (the exact number is the square root of 2). You can prove this using the Pythagorean theorem, and this is shown in Figure 1.2 when you assign the length of the two equal sides as x. This is an elementary example of pure mathematics.

As for applied mathematics, such trigonometric relationships are readily applicable to the real world. Back to gardening, let's say you have a square garden that measures 20 feet on each side and you need to know the length from one corner to its opposite corner (the diagonal length). Using the mathematics described here, you can approximate this length as about 14 feet.

When thinking of the two broad branches of mathematics, applied mathematics may seem the less daunting of the two. The work of pure mathematics involves the invention of new material, whereas applied math takes these efforts to solve new problems. However, this perception is not at all the case, as applied mathematicians have their work cut out in dealing with the "messy" real world. To solve problems, they need to appropriate mathematical ideas that have been created in an ideal, perfect world. It is also true that venturing into the new frontiers of pure mathematics is highly daunting, and many work tirelessly for years at this. This discussion between the two branches suggests that mathematics teaching must include both. Many attempts have been made to "make mathematics relevant" with the inclusion of applied mathematics. Some of these are more contrived (think of those textbook word problems that do not resemble real situations),

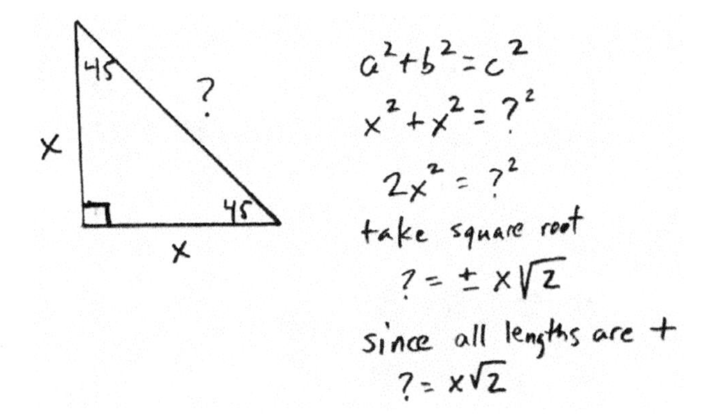

FIGURE 1.2 An elementary example of pure mathematics

and there are other examples that reflect the work of applied mathematics more accurately. As for pure mathematics, teaching mathematics in this way implies that we provide experiences through which our students will come to "invent" mathematical ideas as pure mathematicians do.

For the latter, here is an example adapted from Paul Lockhart's (2009) *Mathematician's Lament: How School Cheats Us Out of Our Most Fascinating and Imaginative Art Form*. We will take a deeper look at this book in Chapter 2, but for now, it helps us explore pure mathematics a bit more with an example problem coming from the mathematical branch of number theory. The most helpful part of Lockhart's example is the fact that he encourages us to imagine any number as a pile of rocks. So, to begin: imagine the numbers 9 and 4 as two separate piles of rocks. Now consider arranging them in various ways, like a line, a circle, a square, more than one object, and so on. It could be helpful for you to get out some beans or something else to use as your "rocks."

I assume you know that 9 is an odd number and 4 is an even number. Continue playing around, arranging your rocks, and this time focus on these facts about even and odd. Is there a way you can arrange the 4 to show that it's even? Maybe create a set of 6 rocks and 8 rocks as well; then you have a few even numbers to play with. Do the same for odd. Come up with as many arrangements as possible. The longer you play, the more likely you are to come up with the arrangement I hope you do. This arrangement, Figure 1.3, appears below. Don't peek until you've played enough!

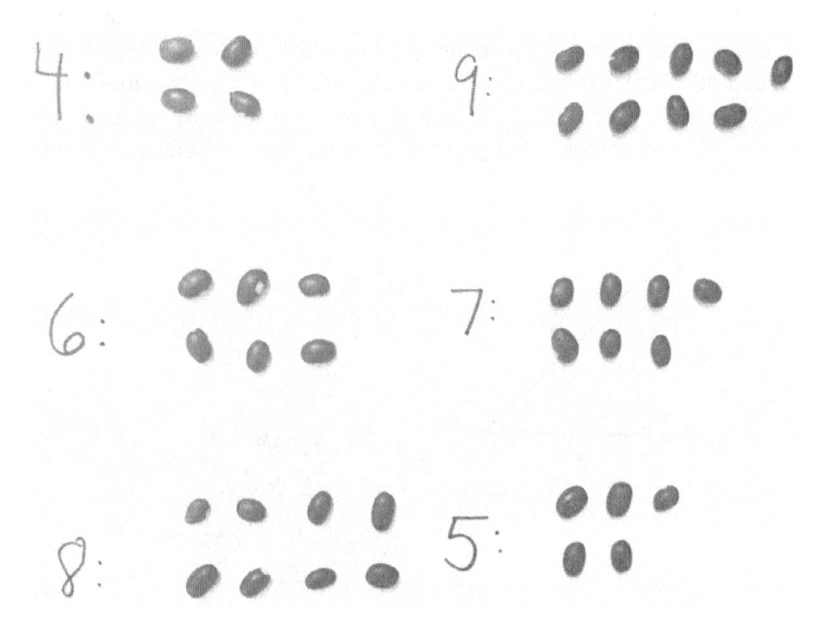

FIGURE 1.3 Beans arranged to deduce theorems from number theory

Any even number can be placed in two equal rows. In other words, we can pair up each of the rocks. If we try to place an odd number of rocks into two rows, one of the rocks is left without a pair. Fascinating!

This representation can be used to answer some interesting questions in number theory. Use these representations (numbers as rocks) to prove an answer to the following questions: What kind of a number do you get when you add two even numbers? Two odd numbers? An even and an odd number? Enjoy playing with these representations and work on a problem in pure mathematics!

There is one final note about the notions of pure and applied math. As with most dichotomies, I suggest you consider them as useful categories to further appreciate the nuances of mathematics. In doing so, however, we cannot come to understand them in any way as distinct entities. There is much of mathematics that may fall into one or the other category. Similarly, the notion of application seems to suggest that pure math always comes first. And, by further logic, that pure math is somehow superior to applied math. Hersh (1997) notes that pure mathematicians value applied math just as highly as pure math. Part of this is the fact that much of pure math has occurred as the result of applied math. "Not only did the same great mathematicians do both pure and applied mathematics, their pure and applied work often fertilized each other. This was explicit in Gauss and Poincaré" (p. 26).

What Does the Philosophy of Math Tell Us?

Exploring these various branches of academic mathematics has begun to answer our central goal in this section. After such an introduction, it seems appropriate to next take a look at the work of philosophers of mathematics. We might suspect such work aims to answer the question in a very direct way. Instead, learning the mainstream and contrary viewpoints within the philosophy of math presents an important consideration you may not have anticipated. Namely, the following review asks us to decide whether mathematics exists outside of our having discovered it, as a set of ideals, or as something that was created by people. The latter represents some of the more controversial and critical aspects to thinking about what mathematics is.

Two books provide a highly comprehensive review of these areas: Hersh (1997), *What Is Mathematics, Really?*, and Ernest (1990), *The Philosophy of Mathematics Education*. Both provide a significant review of the major names in philosophy of mathematics, with Hersh as a narrative style and Ernest as an in-depth and technical review of each strand of philosophies of mathematics. Both pay equal attention to what we might call a mainstream philosophy of mathematics and more critical viewpoints. For those in the mainstream viewpoint, "mathematics is superhuman—abstract, ideal, infallible, eternal." In several cases these thinkers are "tangled with religion and theology" (Hersh, 1997, p. 92). On the other hand, those with the contrary viewpoint see mathematics as a human activity or human creation. Hersh refers to this group as the humanists, and he identifies thinkers on both sides throughout the western history of the philosophy of mathematics. As an example of humanist mathematics philosophy and a philosopher of mathematics himself, Ernest's specific critical viewpoint is termed "social constructivism."

The mainstream view of the philosophy of mathematics is clearly exhibited by ancient Greek philosophers, including the Pythagorean society and Plato. The Pythagorean society situated its mathematical activity within a quest for spirituality. For example,

> The Pythagorean discovery that the harmonics of music were mathematical, that harmonious tones were produced by strings whose measurements were determined by simple numerical ratios, was regarded as a religious revelation.... The Pythagoreans believed that the universe in its entirety, especially the heavens, was ordered according to esoteric principles of harmony, mathematical configurations that expressed a celestial music. To understand mathematics was to have found the key to the divine creative wisdom.
>
> *(quote attributed to Richard Tarnas in Hersh, 1997, p. 93)*

The perfect, ideal relationships witnessed in music and elsewhere indicated a harmonious truth and beauty. It was as if to say, to lead fully spiritual lives, to become more beautiful and perfect, we as people must learn mathematics and discover such harmonies in their existence.

This is regarded as a stepping stone toward Plato's famous notion of ideals, in which mathematics played a significant role.

> Platonism is the view that the objects of mathematics have a real, objective existence in some ideal realm. It originates with Plato, and can be discerned in the writings of the logicists Frege and Russell, and includes Cantor, Bernays, Godel and Hardy among its distinguished supporters. Platonists maintain that the objects and structures of mathematics have a real existence independent of humanity, and that doing mathematics is the process of discovering their pre-existing relationships. According to Platonism mathematical knowledge consists of descriptions of these objects and the relationships and structures connecting them.
>
> *(Ernest, 1991, p. 29)*

Such Platonism is one of a few varieties in the mainstream view of the philosophy of mathematics that imagines mathematics as fixed, neutral, and value free. Another strand of mathematics philosophies, absolutism, includes intuitionism, formalism, and logicism. Logicism describes a standpoint in which all of mathematics can be described within logical terms and principles; formalism essentially claims mathematics to be the practice of defining mathematical truths through symbols; and intuitionism holds that mathematics must rely on the construction of proofs and objects (Ernest, 1991, pp. 7–12). As Ernest (1990) writes, absolutist schools of philosophy should have

> included accounting for the nature of mathematics, including external social and historical factors, such as the utility of mathematics, and its genesis. Because of their narrow, exclusively internal preoccupations, these schools have made no contribution to a broadly conceived account of mathematics.
> *(pp. 23–29)*

Hersh points out that in many instances, such absolutist and Platonic frames of mind coincide with a religious or theological perspective. For example, Descartes attempts to prove that God exists because a perfect triangle exists within his mind. In this way, mathematics is seen as a set of divine ideals to be discovered by people.

On the other hand, a major branch of philosophy of mathematics is termed fallibilism: the view that "mathematical truth is fallible and corrigible, and can never be regarded as beyond revision and correction" (Ernest, 1991, p. 18). If Platonism and absolutism rest on mathematics' attempts to discover what is indubitable, fallibilism represents a mathematical body of knowledge that we know to be true simply because we have not proven it false yet. More broadly, this description fits under what Hersh terms a "humanist" mathematics, in which mathematics is seen as the product of human interaction and contestation. Ultimately, any mathematical truth has been argued by people and is thus the product of such human experiences. Among many more, two philosophers of mathematics are important here: Ludwig Wittgenstein and Imre Lakatos. Wittgenstein declares the following: "1 plus 1 equals 2 because we have decided it so." This is an important step and a clear disagreement with the mainstream absolutist and Platonic perspectives. The sum of two 1s equals 2 not because of some ideal set of numbers, existing in their perfect form and perhaps with divine intervention. Instead, over the course of human history, we have decided it so. Any person is free to disagree with the equation. For example, they might say that the sum is 3. However, this person would have difficulty participating in the mainstream use of numbers that society has developed over time.

Lakatos described the process in more detail. Influenced by Karl Popper, a philosopher of science, essentially he claimed that every mathematical truth is the result of argumentation. Popper had revolutionized science by arguing that scientific theories are only guesses waiting to be disproved by experimentation. Similarly, mathematical truths are statements either proven true via an argument that is accepted by the mathematical community (a "proof") or proven false given a "counterexample" or perhaps other means. Earlier, we played with the notion of proof when attempting to explain that the sum of two even numbers is even earlier in this chapter. For the record, mathematicians would not accept our play with rocks as a formal proof, but for conversation's sake this will be helpful. How about the counterclaim that two even numbers add to an odd number? Well, that is easy to argue against with the use of a counterexample. Take a minute to give one now by playing with your rocks, beans, and so forth.

Proving and providing counterexamples leads to mathematical "truth." Hersh writes,

> Instead of a general system starting from first principles, Lakatos presents clashing views, arguments, and counter-arguments; instead of fossilized mathematics, mathematics grows unpredictably out of a problem and a conjecture. In the heat of debate and disagreement, a theory takes shape. Doubt gives way to certainty, then to new doubt.
>
> *(Hersh, 1997, p. 211)*

Embedded within Lakatos's assertions is the assumption that mathematics is not ideal truth and certainly not something created by God and that we discovered. This point clashes somewhat with the mainstream philosophies of math. However, I suggest it does not contradict the mathematical quest of discovering truth and beauty, such as exemplified by the Pythagoreans and their attempts to live the good life. What remains important in our objectives for studying mathematics education critically is that we examine the variety of natures of mathematics expressed by this philosophical work.

A specific option available to us is Ernest's own philosophy of mathematics, what he terms the social constructivist position. Hersh positions Ernest's work within a humanist philosophy of mathematics. Ernest begins with the clear statement that mathematics is a "social construction." This concept will be utilized consistently throughout this book, as we look at other examples of social constructs such as race, class, and gender. For now, think about social constructs as objects that we might think of as "fixed realities" but instead have been developed over time in social settings. Social theorist Michel Foucault uses the term "regimes of truth." Do we not usually think of mathematics as fixed and objective? True and value free? And certainly the earlier philosophies of mathematics, like absolutism, reinforce this assumption. Alternatively, we can think of mathematics as something manufactured by social groups, a social construct.

To begin, Ernest (1991) orients us to a claim that mathematics is delivered via language, which on its own is a construction of the social experience. In describing the process in which mathematical knowledge comes to be, Ernest distinguishes between objective and subjective mathematical knowledge. An individual constructs subjective mathematical knowledge, and objective mathematical knowledge is that which has been understood and accepted by a community of mathematical knowers. When an individual proposes a new mathematical statement, she uses language. The community of mathematical knowers uses objective mathematical knowledge to make sense of this new statement. The body of objective mathematical knowledge is the discipline of mathematics, and such a conception fits within a fallibilistic claim that everything known mathematically has simply not yet been disproven.

Ernest stops short of discussing this community of mathematical knowers. He claims that a philosophical inquiry can only go so far as to describe the generation of mathematical knowledge through this social process.

> It would be inappropriate in a philosophical account to specify any social groups or social dynamics, even as they impinge upon the acceptance of objective knowledge. For this is the business of history and sociology, and in particular, the history of mathematics and the sociology of its knowledge.
>
> *(Ernest, 1991, p. 63)*

That said, Ernest's philosophy of social constructivism begs us to ask these questions as we go about our critical understanding of mathematics. We have thus opened the door to our final two sections of this chapter: looking to the history of mathematics and the field of ethnomathematics. It is my hope that these sections will help you more fully appreciate how mathematics is a social construct. As you read, consider how these contributions describe a social process by which mathematical knowledge has been created.

What Does the History of Math Tell Us?

History of mathematics is an important field that, on its own, helps us address the question of what mathematics is. It will help us to think more deeply about the social process by which mathematical knowledge has developed. Reading these histories reveals to us an essential feature of this domain, namely that historians of mathematics are either entirely concerned with western mathematics or with the mathematics of "other" cultures. By drawing attention to this point, we come to know not only what is math but also *who* does mathematics, thus broadening our inquiry in the chapter. For example, learning the history of mathematics indicates that the west adopted many mathematical practices from the east, like arithmetical notation and trigonometry, and that Mesoamerican people (e.g., Maya, Aztec) developed sophisticated mathematical explanations of astronomical patterns. Another point we will consider is the set of particular topics within the entire domain mathematical knowledge, with special attention to statistics.

Specific works that are designated as histories of mathematics typically present an overview of the progression of western mathematics. For example, Roger Cooke's (1997) *The History of Mathematics: A Brief Course* is broken down into three sections: modern western mathematics (developments since the Middle Ages) and its two influences, early western mathematics (ancient Greece through the Roman Empire), and nonwestern mathematics (such as Chinese, Hindu, and Muslim societies). In this way, this body of work focuses on western mathematics.

It also implies a lack of mathematical invention in a host of other societies, such as indigenous societies of sub-Saharan Africa and North and South America. For these contributions, we will turn to the field of ethnomathematics in the next section.

To start, histories of mathematics suggest the importance of a handful of ancient societies that contributed to modern western mathematics. These include Egyptian, Babylonian, Greek, Roman, Hindu, Chinese, Japanese, Korean, and Muslim people. All made contributions to a variety of branches of mathematics that were further developed in the modern period. These groups developed such topics as computation, number theory, geometry, algebra, and applied mathematics. The following are some examples of these contributions. Read these examples to experience the diversity in influences on modern western mathematics as well as to further your explorations of the concepts within mathematics.

Early mathematical practices existing in India and China are now very typical practices across the globe. "Decimal notation and the symbols for numerals we use today originated in India and came to Europe through the Arabs" (Cooke, 1997, p. 197). Over time and across societies, the format for each symbol representing the digits has changed. That is, there have been many different symbols to represent 1, 2, 3, . . . In particular, then, the Indian influence was the practice of using a single symbol for 10 digits and then using these to describe any number as a sequence of digits. This saves us from continuing to invent new symbols for the various numbers. For example, we could have: 1, 2, 3, 4, 5, 6, 7, 8, 9, &, #, @, where & is a symbol to mean 10, # means 11, and @ means 12. Or we could have our number 10, which relies on the concept of *place value* and strings two symbols together. The number 427 indicates that there are 4 hundreds, 2 tens, and 7 units. We can thus describe any number with only 10 symbols and an understanding of place-value notation. The place-value practice in base 10 was used both by the Indians and Chinese at the time. It has been difficult to determine who used it first, if that might be your interest, but

> It certainly came to the West from the Arabs, who learned it from India. In fact, one of the influential treatises by which Europeans learned about the decimal system and the symbols for digits was a treatise by the Muslim scholar Kushyar ibn Labban.
>
> *(Cooke, 1997, p. 197)*

Our numeral system is often referred to as Arabic, but it has also been referred to by other names to reflect more accurate historical understandings, such as Hindu-Arabic numerals.

Indian influence also included their dealings with number theory, or the branch of mathematics that studies whole numbers and rational numbers. Typical problems in the field include finding prime numbers and divisibility. For one, Hindus were interested in the triples of integers for which the sum of the squares of the two smaller equals the sum of the square of the larger. Two examples of these triples are the numbers (3, 4, 5) and (9, 40, 41). You may have encountered these before under the name "Pythagorean triples." While the Pythagoreans may have been interested in their practical use as related to right triangles, it is possible the Hindus found a religious purpose to this project:

> A Hindu home was required to have three fires burning at three different altars. The three altars were to be of different shapes, but all three were to have the same area. These conditions led to certain "Diophantine" problems, a particular case of which is the generation of Pythagorean triples, so as to make one square integer equal to the sum of two others.
>
> *(Cooke, 1997, p. 198)*

This example shows how the context within which mathematical knowledge originates can be surprising. Perhaps we might expect to have found the first use of Pythagorean triples in a topic more relevant to engineering. The religious nature to this origination serves as an example of the socially constructed nature of mathematics. Recall Ernest's notion of the social constructed nature to mathematics. In this example from Hindu culture, the community of knowers validated the mathematical knowledge because it fulfilled a particular desire in a religious context.

Similar to number theory, algebra emerged among a variety of locations and cultures. Its title comes from the Arabic word *al-jabr*, used by Muhammad ibn Musa Al-Khwarizmi of the ninth century. His work centers on solving equations with an unknown by keeping the equation balanced. This can relate to common practices in mathematics classrooms and modern algebra. For example, to solve the equation $3x + 9 = 12$, we can first subtract 9 from both sides to keep the equation balanced. In this way, Al-Khwarizmi was interested in developing an algorithm, or procedure, for solving equations with unknowns. This goal came about as the result of extensive work in dealing with equations with such unknowns. Many consider this the essential feature of elementary algebra: solving equations to find an unknown value.

With this quest to find unknowns as the focus of algebra, most consider the "father of algebra" a toss-up between Al-Khwarizmi and the Roman mathematician Diophantus of Alexandria. Likely written in the second century C.E., his *Arithmetike* contributes several practices that are commonly used in algebra. As Roger Cooke notes, these include using symbols to represent an unknown number (like using x in the earlier equation) and describing such an unknown so that it can be determined. For the same equation ($3x + 9 = 12$), we could give the following description: "I am thinking of a number. When you multiply this number by 3 and add 9, you get 12. Can you tell me the number?" This is a game that some algebra teachers use with their students to begin understanding the idea of an unknown number, as well as fostering the students' development of an algorithm to answer it. Both Diophantus and Al-Khwarizmi focused on the ways to find unknown values and especially attempted to develop algorithms that do this.

These examples from the history of mathematics aim to decenter a myth that modern mathematics is a western conception. If anything, the pattern seems to be the appropriation of eastern concepts by western civilizations. Furthermore, the development of mathematics is rich with social contexts, and such histories help us consider mathematics as a construction by groups embedded in social life. This answers the questions put forth by Ernest's philosophy of mathematics termed social constructivism, introduced previously. Individual contributors to mathematical knowledge are not disconnected from the context of social life. They aim to invent concepts and communicate these through a language that will be accepted by the mathematical community. Mathematical knowledge is also embedded with other aspects of social life, including practical matters like engineering as well as spiritual matters.

Ethnomathematics: Thoughtfully Considering an Anthropology of Mathematical Knowledge

Having reviewed some examples from the history of mathematics, we move to the anthropologies of mathematics that further expand our conception of mathematics. This work, often referred to as ethnomathematics, suggests several important points. First, it reminds us that the roots of mathematical knowledge are computation, arithmetic, and geometry. It also suggests a novel answer to our main question that echoes Ernest's social constructivism: *mathematics is a language*. Finally, ethnomathematics continues to reject mathematics as a uniquely western project, as the history examples have done.

To begin, we should review just what is meant by the term. A major figure in ethnomathematics is Ubiritan D'Ambrosio. His (2002) survey of the field proves an excellent source to grasp its orientations and dispositions. As he puts it,

> Ethnomathematics is the mathematics practiced by cultural groups, such as urban and rural communities, groups of workers, professional classes, children in a given age group, indigenous societies, and so many other groups that are identified by the objectives and traditions common to these groups. *(p. 1)*

This claims that mathematics exists in a multiplicity of practices and, a point relevant to our teaching, that students and communities have mathematics embedded in their lives. Thus D'Ambrosio suggests that ethnomathematics is essential to best practices in pedagogy.

One mathematics education scholar, Alexander Pais (2011), cautions us to think more carefully about these applications of ethnomathematics. He provides examples of when students have participated in a lesson in ethnomathematics, only to walk away with an unchallenged viewpoint of social relations. For them, the mathematical practices of "Other" cultures and people continue to be objectified. Pais also encourages the work of ethnomathematicians to focus more on the cultural practices of academic mathematicians themselves. This fits within the stated objectives for ethnomathematics laid out by D'Ambrosio, but as Pais points out, all too often, the field focuses more on seeking out the mathematics of the "Other" rather than understanding more fully how mathematical knowledge comes to be socially constructed by academic mathematics. All in all, these efforts can provide more depth to the philosophical points made by Ernest regarding mathematics as a social construction.

Nevertheless, the contributions from ethnomathematics decenter our fixed conception of mathematics and deserve our attention. Several examples are contained in the edited volume *Mathematics Across Cultures: The History of Nonwestern Mathematics*. One author in this volume characterizes the field of ethnomathematics as having two branches: the first is a "general anthropology of mathematical thought and practice" in every geographic area, and the second is the specific dedication to understanding the mathematical practices of small scale and indigenous cultures (Eglash, 2000, p. 13). We have already uncovered much of the work in the former. For example, historical and anthropological inquiries have helped us understand how the most widely used numeral system came from eastern societies. In other words, the first definition of ethnomathematics is in looking for the nonwestern influences on modern western mathematics. In contrast, the second definition does not concern itself with questions of influence and instead is determined to validate other cultures' use of mathematics. This project addresses the implication that mathematical behavior exists only in particular societies, such as those with written language, urban centers, agriculture, and/or hierarchical state structures.

Here are some examples of the findings in this second branch of ethnomathematics. The Incan use of quipus has been well documented. These knotted ropes maintained records of transactions and tax collection, and their patterns indicate significant mathematical computation beyond simple arithmetic. Mayan culture engaged significantly with mathematics: Some paintings on pottery illustrate their math classrooms; they also had an understanding of zero, which you may not realize to be a major development in mathematical thinking.

Mathematical practices are also embedded in cultural practices, be they artistic, religious, or practical. People living in the Great Plains of pre-Columbian North America constructed tipis, an architectural structure with several mathematical properties. Similarly, African-American quilts that preserve personal histories via the use of colors, beads, and knots all have significant mathematics underlying them. Some might suggest we proceed with caution as we continue with such examples. Describing the mathematics of the tipi cone uses western understandings and commits to the anthropological "gaze." That is, we are searching for western things embedded in indigenous cultures. This is a project to elevate these cultures, because, previous to such discovery, European people thought there was nothing to be valued in them. Instead, we can use our western understandings as an entry point into the thinking of indigenous cultures, which will in turn expose new ideas about human behaviors in all its varieties. In this case, we are thinking about these examples so that our understanding of what mathematics is can be complemented by other behaviors. In addition to mathematical operations and the various fields within mathematics, we might add quilt making to our list of mathematical behaviors.

Ethnomathematics provides other contributions to our understanding of the nature of mathematics, such as the suggestion that math is a language:

> Mathematics is a method for communicating ideas between people about concepts such as numbers, space and time. In any culture there is a common, structured system for such communication, whether it be in unwritten or written forms. These systems can form bridges of communication across culture and across time . . . Mathematics has been part of all societies, a part of every profession as well as everyday life. Western mathematics became narrower with the insistence that only deductive mathematics from a set of axioms, following the Greek tradition, was *real* mathematics . . . [And,] mathematics has often worked on many levels, as part of everyday culture and also as used by subgroups within the main culture . . . Indeed in many cultures, the mathematics of calendars and astronomy were in the hands of the priestly classes.
>
> *(Wood, 2000, pp. 1–3)*

We see mathematical communication among all cultures and, interestingly, several cultures aimed to preserve some mathematical knowledge for a subset of their populations. By viewing mathematics as a language, we also realize that mathematics exists anywhere there is communication, including cultures without written language. Mathematical communication does not have to be written down, as evidenced by the intricate counting systems of Papua New Guinea and Oceania. Other examples are the weaving patterns of Northern Australian aboriginals and the knotting of quipus as discussed earlier. As we begin to shift our attention toward education, we can look at how communications can be mathematical. Communications include written, oral, and visual, through artwork or other representations, for example.

With our attention to ethnomathematics in this section, we have come to understand the major contributions as well as a caution to its pitfalls. It is important to remember that through ethnomathematics we come to fully appreciate what can be considered mathematical behavior. However, we need to think carefully about whether such thinking further objectifies "Other" cultures by studying them through western eyes and, importantly as we transition to discussions of mathematics education, how we bring these discussions into the classroom.

In summary of this chapter, we have covered quite a bit of territory that can help us understand what mathematics is. The introduction of mathematical operations and academic mathematics was a natural place to start, given that this is likely what we all think of mathematics being in the first place. However, we next moved into some of the contests within the philosophy of mathematics and have put to you the question of whether mathematics exists without humans or was in fact constructed by them. A more critical viewpoint embraces Ernest's "social constructivism." Here, mathematical knowledge is the product of the mathematical community, itself embedded within historical and social contexts. It is worth mentioning here that Chapter 4, on gender and mathematics education, will provide a feminist interpretation of mathematics along these lines. For now, it is important to stress that these considerations from the histories of mathematics and the field of ethnomathematics decentered mathematics as an exclusively western project. They also (along with the philosophy of mathematics exploration) emphasized the linguistic nature of mathematics, in which mathematical ideas are primarily communications between people. All of these considerations are highly relevant to teaching mathematics with a critical perspective.

At the conclusion of each chapter in this book, I provide activities and prompts for your consideration that expand on the contents discussed. You are encouraged to work collaboratively on these as well as to take a look at the suggested readings from which I sketched this review of mathematics.

Activities and Prompts for Your Consideration

1 Mathematics is typically viewed as an objective, value-free knowledge. For example, it is hard to dispute that 2 + 2 equals 4, correct? From your own experiences in mathematics and this chapter's sketch, how do you respond to the notion that mathematical practice is "black and white"? Do you side with a more absolutist or social constructivist philosophy of mathematics?
2 Browse some of the projects from Ethnomathematics on the Web, a collection maintained by Ron Eglash (link: http://isgem.rpi.edu/pl/ethnomathematics-web). As you browse, keep in mind a few of the thoughts about ethnomathematics discussed in the chapter. Try to find an example that demonstrates the linguistic nature of mathematical practice. Can you find an example that does a good job providing a rich cultural context to the mathematical practice? On the other hand, did you find any examples that objectify a cultural practice by simply "finding the western mathematics" contained in it? How might you incorporate one of these projects into your teaching of mathematics?

3 Browse the website http://www.storyofmathematics.com/ which presents a history of mathematics. Does this history as presented perpetuate a western-centered view of the history of mathematics? Does the website seem to align with a particular nature or philosophy of mathematics, such as those described in this chapter?

References

Cooke, R. (1997). *The history of mathematics: A brief course.* Hoboken, NJ: Wiley.

D'Ambrosio, U. (2002). *Ethnomathematics.* Rotterdam, The Netherlands: Sense.

Eglash, R. (2000). Anthropological perspectives on ethnomathematics. In *Mathematics across cultures: The history of non-western mathematics*, edited by H. Selin. Dordrecht, The Netherlands: Springer, pp. 13–22.

Ernest, P. (1991). *The philosophy of mathematics education.* New York: Routledge.

Hersh, R. (1997). *What is mathematics, really?* New York: Oxford University Press.

Lockhart, P. (2009). *A mathematician's lament: How school cheats us out of our most fascinating and imaginative art form.* New York: Bellevue Literary Press.

Pais, A. (2011). Criticisms and contradictions of ethnomathematics. *Educational Studies in Mathematics* 76: 209–230.

Wood, L.N. (2000). Communicating mathematics across culture and time. In *Mathematics across cultures: The history of non-western mathematics*, edited by H. Selin. Dordrecht, the Netherlands: Springer, pp. 1–12.

2

INITIAL EXAMINATIONS OF MATHEMATICS EDUCATION

Purpose, Problems, and Method

With the preceding exploration of mathematics at hand, we now turn to our work that occupies the remainder of this book: critically examining mathematics education. This chapter introduces these efforts first with a review of four books from popular mathematics education literature that question the ways mathematics education is typically conceived. These will introduce the tenets central to reform mathematics education, the first step in teaching mathematics with a critical perspective. In the second half of the chapter, we look at the modern practice of teaching mathematics within a typical school. This acknowledges the fact that many of you work or will work within this space. Here our goal will be to examine school mathematics' typical structure, meaning the lesson plan, deemed so essential to classroom mathematics teaching. This will allow us to consider how we can incorporate the critical perspectives contained throughout this book into your teaching practice.

Popular Introductions to Mathematics Education

To begin, we review four works that, in my view, initiate plentiful discussion about mathematics education and open up questions regarding the traditional methods by which it is taught. By no means do I suggest that these are the four most critical books on mathematics education. Rather, these present some of the more popular criticisms that open us to thinking more deeply about mathematics education. These books help us start questioning *how* mathematics should be taught, *to whom* we teach it, and *for what reason*. The conversations in this review open these questions up for full exploration, and these will remain in play in subsequent chapters. Primarily, they will introduce reform mathematics teaching with its focus on mathematical thinking and conceptual understanding, as well as introducing the notion of mathematics for all, or equitably teaching mathematics.

The first book is *Knowing and Teaching Elementary Mathematics: Teachers' Understanding of Fundamental Mathematics in China and the United States* by Liping Ma (2010). This highly celebrated book discusses in depth a major trend in mathematics education research that emphasizes a teacher's content knowledge and pedagogical content knowledge. Ma researched the mathematical content knowledge and pedagogical content knowledge of elementary teachers in both China and the United States to indicate that China's teachers had greater command of mathematical knowledge and how to make this accessible for students. Her research was motivated by what she describes as a paradox, in which the Chinese teachers are less educated in terms of years and yet the Chinese students fare better on international comparison tests. In her book, she suggests an explanation for the paradox:

> My data suggests that Chinese teachers begin their teaching careers with a better understanding of elementary mathematics than that of most U.S. elementary teachers. Their understanding of the mathematics they teach and—equally important—of the ways that elementary mathematics can be presented to students continues to grow throughout their professional lives. Indeed, about 10% of those Chinese teachers, despite their lack of formal education, display a depth of understanding which is extraordinarily rare in the United States.
>
> *(Ma, 2010, p. xxvi)*

Ma's research requires careful consideration when teaching mathematics with a critical perspective. I introduce two of these now and explore them in the next few paragraphs. First, it prioritizes a mathematics teaching and learning in which the mathematical material is made accessible to the students and in which conceptual understanding and development are important for the learning process. This is a trend in mathematics education that we will see in a few of the other books. Second, Ma's study is emblematic of what might be termed apolitical mathematics education scholarship. Here, the goals of mathematics education are assumed and unstated, and the research is framed by acceptance of status quo practices, such as the validity and need for international comparisons of mathematics education. As we take these two points in more detail, keep in mind our present goal of introducing the landscape of mathematics education via a critical discussion.

Mathematics education scholarship has for a long time pushed against traditional rote learning of mathematical skills and concepts and represents the first steps in teaching mathematics critically. The traditional view promotes drill, memorization, and sequential learning through a set of topics that, once complete, will lead to conceptual understanding. This view is still held by many practicing mathematics teachers and promoted by a significant number of famous mathematicians, including several involved in mathematics policy. However, the dominant view in mathematics education research promotes an alternative viewpoint in which students learn mathematics by *doing it*. They are given a rich set of experiences in which to discover, question, problem solve, reason and justify, and communicate their mathematical thinking. Students' deeper understanding is considered relevant when learning new mathematical concepts and skills rather than an afterthought or something that might happen later on. This approach to teaching mathematics has been heralded by mathematics education researchers since at least the 1960s, when, in the U.S., new mathematics curriculum as motivated by the Cold War made these switches. The switch is from direct teaching of math facts toward creating a classroom of mini-mathematicians who think, reason, discover, conjecture, and communicate mathematical ideas. The "new math" as it came to be known spurned a "back-to-basics" movement of the 1970s and again a "new-new math" in the late 1980s. Despite such pendulum swings, the mathematics education research program consistently emphasizes and promotes such reforms, and this can be seen in the publications of the U.S.'s National Council of Teachers of Mathematics (NCTM). I provide a further sketch of this history in the final chapter of this book.

Ma's research on Chinese and U.S. teachers fits within this paradigm of mathematics education research. One of Ma's research prompts for the Chinese and U.S. teachers involved a hypothetical situation in which a student reveals her new theory: that as the perimeter of a rectangle increases, so does its area. The student gives an example of two rectangles that support her theory. In the research prompt, teachers are required to react to the situation and develop a response to the student. This research prompt is emblematic of reform mathematics teaching because it emphasizes conceptual understanding of the mathematics and suggests that students should be asking these types of questions and should be pushed to investigate them on their own. In a traditional view, a teacher would simply state that the student's theory is incorrect and get back to practicing with the area and perimeter formulas. However, reform mathematics pushes students and teachers to behave more like mathematicians than robots. In comparing the U.S. and Chinese teachers, Ma finds the U.S. teachers to be competent with the procedures of mathematics but lacking in the mathematical process:

> The U.S. teachers did not show glaring weaknesses in their calculation of perimeter and area of rectangles. However, there was still a remarkable difference between the U.S. teachers and their Chinese counterparts. Only three U.S. teachers (13%) conducted mathematical investigations on their own and only one reached a correct answer. On the other hand, 66 Chinese teachers (92%) conducted mathematical investigations and 44 (62%) reached a correct answer. Two main factors may have precluded the U.S. teachers from a successful mathematical investigation—their lack of computational proficiency and their layperson-like attitude toward mathematics. Although

> most of the U.S. teachers knew how to calculate the two measures, they were far less proficient than their Chinese counterparts. A few reported that although they could do the calculations, they did not understand their rationales, and that this deficiency hampered further exploration. This was not the case for the Chinese teachers. None reported that lack of knowledge about the formulas hindered their investigations. The second factor, which may be even more significant, was the teachers' attitudes toward mathematics. In responding to the student's novel claim about the relationship between perimeter and area, the U.S. teachers behaved more like laypeople, while the Chinese teachers behaved more like mathematicians. This difference displayed their different attitudes toward mathematics.
>
> *(Ma, 2010, p. 104)*

Notice in the previous quotation how the language of mathematics education prioritizes conceptual understanding, reasoning, and other skills of mathematicians. On this question, Ma's work clearly stands on the reform side of mathematics education.

On the other hand, *Knowing and Teaching Elementary Mathematics* confirms traditionalist perspectives on the mathematics education project in total, albeit by omission. This is also typical of reform mathematics education work and a topic I discuss at length in my book *Math Education for America? Policy Networks, Big Business, and Pedagogy Wars* (2014). Reform mathematics, as well as traditional mathematics, fails to take a critical perspective on the purpose for mathematics teaching and learning. If they state it at all, both sides of the math wars (the majority of traditionalists and reformers) claim mathematics education as essential for economic growth, a human capital viewpoint in which schools are seen as economic investments that develop natural resources, in this case people, from which companies can gain profit. As is often the case, Ma does not state this orientation outright; rather, her commitment lies in the framing of her study. She takes at face value the international comparison test scores without question. They motivate her study but at the same time firmly commit to human capital motives for education. Furthermore, we can be critical of Ma's study for its classification of people as "us" and "them," thereby further reinforcing racial and other categorical stereotypes and assumptions. While her work points us toward what reform mathematics teaching can be about, it does little to help us think about the purposes of mathematics education and how it can move us toward social justice and sustainability.

Ma's work is but one among many that orient us toward a mathematics education that opens minds to inquiry, reason, communication, and the other facets of reform mathematics teaching and learning. This represents the first step in teaching mathematics with a critical perspective because it resonates with a greater variety of the philosophies of mathematics covered in the previous chapter. As we did there, critically thinking about mathematics reveals the assumptions we usually hold: that mathematics is static, neutral, value free. Such a conception of mathematics, the absolutist view, engenders a traditional, direct-instruction approach to teaching it. The data on success in mathematics, however, indicates that many cannot benefit from such a style of teaching, and the socially constructed nature of mathematics, as put forth by Ernest, begs us to think differently about how to make it accessible for more students.

Another champion of the reform era in mathematics education is Jo Boaler. Her accessible publication *What's Math Got to Do With It? How Parents and Teachers Can Help Children Learn to Love Their Least Favorite Subject* (2008) lays out arguments for reform-based mathematics in clear terms. She paints pictures of classrooms like the reform mathematics educators she hopes will exist in widespread numbers. Notice the social activity and meaning making as Boaler describes "teacher Emily's" mathematics instruction:

> All eyes were on the front, and I realized that the students had not heard the door because they were deeply engaged in a problem Emily had sketched on the board. They were working out the time it would take a skateboarder to crash into a padded wall after holding on to a spinning merry-go-round and letting go at a certain point. The problem was complicated, involving high-level mathematics. Nobody had a solution, but various students were offering ideas. After the boys sat down, three girls went to the board and added to the boys' work, taking their ideas further. Ryan, a tall boy sporting a large football ring, was sitting at the back and he asked them, "What was your goal for the end product?" The three girls explained that they were first finding the rate that the skateboarder was traveling. After that, they would try to find the distance from the merry-go-round to the wall. From there things moved quickly and animatedly in the class. Different students went to the board, sometimes in pairs or groups, sometimes alone, to share their ideas. Within ten minutes, the class had solved the problem by drawing from trigonometry and geometry, using similar triangles and tangent lines. The students had worked together like a well-oiled machine, connecting different mathematical ideas as they worked toward a solution. The math was hard and I was impressed.
>
> *(2009, pp. 1–2)*

As Boaler suggests and we all know, this is not the typical mathematics classroom. Mathematics teaching and learning is often uninspired, by the book, and leaving little room for argumentation. Boaler's sketch of mathematics teaching and learning as it could be suggests that students be active, social, engaged, purposeful, and motivated by meaning, just as mathematicians are, as they came to know mathematical properties, concepts and procedures. This sketch does not imply that all such mathematical learning have a practical application, like the skateboarding problem does. The emphasis is on meaning making, and we will see this put a different way when discussing another of the four books in this review.

Reform mathematics teaching: This pedagogy resonates with the social constructivist philosophy of mathematics. It encourages classrooms to model the work of mathematicians, where learners problem solve, discover, conjecture, reason, justify, prove, and communicate mathematical knowledge. Also emphasizes "mathematics for all." The National Council of Teachers of Mathematics (NCTM) and the math education research community promote these teaching methods.

Before leaving Boaler, however, it is important to note that her vision for mathematics teaching and learning continues to commit to the mainstream viewpoint of the purpose for mathematics education. In her introductory chapter, she articulates the importance of learning mathematics as it relates to employer needs. To be fair, she motivates mathematics education for its relevance to both "work and life," the latter describing the necessity to know mathematics so as to function in society. Thus we see that some mathematics educators, in the case of Boaler, for example, do provide greater discussion of the purpose of mathematics education. That said, her uncritical embrace of human capital as the primary motive for mathematics education is typical for most mainstream mathematics educators and is a standpoint less often discussed than the pedagogical debates regarding mathematics teaching and learning. I suggest that such conversation needs to be promoted and thought through if we are to examine mathematics education more critically.

Thus, the most common push against traditional mathematics comes in the form of reform mathematics teaching, also known as constructivist mathematics or, less accurate but more casual, discovery teaching. Very often this gets interpreted as teachers giving a purpose for learning the mathematics by motivating learning through *real-world problems*. The sketch by Jo Boaler is an example. However, reform mathematics teaching more broadly refers to an emphasis on meaning making as opposed to the traditional emphasis on drill and practice divorced from understanding. So, what are other examples of meaning making besides using real-life problems that deepen conceptual understanding? For some answers, we next look to Paul Lockhart's (2009) *A Mathematician's Lament: How School Cheats Us Out of Our Most Fascinating and Imaginative Art Form* because it provides several clear and concise examples of what this looks like.

This little book by Lockhart has a fascinating story. Lockhart, originally a research mathematician who now teaches high school mathematics, wrote the text for this book originally as an unpublished essay that circulated in PDF format among the research mathematics community. Many agreed on its merits and it became a popular text, so much so that Lockhart next published it as a book. Its success is due in part to a clear articulation of what the work of mathematicians looks like and how school mathematics is really nothing like this. The primary message is to reenvision mathematics as an art, and the book opens with an analogy story explaining what music would look like if it were treated as mathematics:

> A musician wakes from a terrible nightmare. In his dream he finds himself in a society where music education has been made mandatory. "We are helping our students become more competitive in an increasingly sound-filled world." Educators, school systems, and the state are put in charge of this vital project. Studies are commissioned, committees are formed, and decisions are made—all without the advice or participation of a single working musician or composer. Since musicians are known to set down their ideas in the form of sheet music, these curious black dots and lines must constitute the "language of music." It is imperative that students become fluent in this language if they are to attain any degree of musical experience; indeed, it would be ludicrous to expect a child to sing a song or play an instrument without having a thorough grounding in music notation and theory. Playing and listening to music, let alone composing an original piece, are considered very advanced topics and are generally put off until college, and more often graduate school.
>
> *(Lockhart, 2009, pp. 15–16)*

The analogy continues, with parallels from the traditional, rote learning and memorization that is typical of mathematics education to the absurd pressures of high-stakes standardized testing.

In making the analogy, Lockhart positions mathematics as an art. "The fact is that there is nothing as dreamy and poetic, nothing as radical, subversive, and psychedelic, as mathematics . . . Mathematics is the purest of the arts, as well as the most misunderstood" (2009, p. 23). I will not go so far as to agree with any superlative declaration regarding mathematics, but the point to take here is that in mathematics, there is meaning making, freedom of expression, creativity, communication, and social activity. Thus Lockhart provides a compelling articulation of much of what we hold true in reform mathematics teaching. As for meaning making, his examples provide illustrations of mathematics teaching and learning that are not bound by applications, like the example we saw in Boaler. In fact, Lockhart is pretty strong in his words against such practice:

> The saddest part of all this 'reform' are the attempts to "make math interesting" and "relevant to kids' lives." You don't need to *make* math interesting—it's already more interesting than we can handle! And the glory of it is its complete *irrelevance* to our lives. That's why it's so fun! . . . In any case, do you really think kids even want something that is relevant to their daily lives? You think something practical like compound

> interest is going to get them excited? People enjoy fantasy, and that is just what mathematics can provide—a relief from daily life, an anodyne to the practical workaday world.
>
> *(2009, p. 38–39)*

It may be wrong to swing completely in this direction and have zero applied problems. For example, how can we account for Boaler's skateboarding problem that motivated students in a real classroom? There seems little reason to focus mathematics education entirely on pure mathematics and leave applied in the dust; however, the reverse is true as well. Unfortunately, the coercive nature that textbooks and curriculum take to making math applicable to students lives does not have the intended effect. And more importantly, we can make mathematics meaningful simply by challenging students to think about the underlying principles that we are teaching them. Lockhart provides a wonderful description (2009, pp. 24–28) of the kind of reasoning that a lesson on the area of a triangle can include. Instead of a mathematics learning that emphasizes the use of the formula, students will derive it themselves and then know it for life. Reform mathematics teaching does not discount the need for the practice of mathematics as well, so rest assured that reform teaching will ensure that students can *use* the formula for the area of a triangle as well!

I have reviewed these first three books to give some substance to the various points within reform mathematics teaching. Reviewing these three books represents how reform mathematics teaching encompasses a broad range of perspectives. While some may be more or less critical in a broader sense, all three push us to think about how to teach mathematics with meaning. This is the important first step in teaching and learning mathematics with a critical perspective. A traditional perspective on mathematics teaching and learning limits students to rote learning and does not allow them to question. By emphasizing meaning making, practical application, social activity, and reasoning, we are increasing access to mathematics for more learners. On this point of increasing access, we turn to the last book in this introductory review of the more mainstream viewpoints on teaching mathematics more critically.

Robert Moses is a civil rights activist who began his work in the 1960s at the height of the movement, helping register voters in the U.S. South. Nowadays, he continues the work of civil rights with a focus on education and, in particular, refers to mathematics education as the "new civil right." In his book *Radical Equations*, Moses and Cobb (2001), he describes mathematics and algebra as gatekeepers to economic opportunity. In so doing, he draws upon the demographic statistics of math-intensive careers as well as college graduation requirements that emphasize algebra. Mathematics education is thus viewed as a gatekeeper, preventing many working-class and people of color from attaining college degrees and economic success. In response, for more than 10 years Moses has developed and sustains his "Algebra Project" to increase access to mathematics to African American and Latino/Latina students in the United States:

> Organizing around algebra has the potential to open a doorway that's been locked. Math literacy and economic access are the Algebra Project's foci for giving hope to the young generation. That's a new problem for educators. It's a new problem for the country. The traditional role of science and math education has been to train an elite, create a priesthood, find a few bright students and bring them into university research. It hasn't been a literacy effort. We are putting literacy, math literacy, on the table. Instead of weeding all but the best students out of advanced math, schools must commit to everyone gaining this literacy as they have committed to everyone having a reading-writing literacy.
>
> *(2001, pp. 16–17)*

In these efforts, the Algebra Project is as much about new teaching strategies as it is about community organizing to demand better educational opportunities and increase awareness among the community. And to be sure, these teaching strategies resonate with the reform approaches we covered earlier.

In looking at these four popular books, we come to see the most popular criticisms of mathematics education and how to begin to think differently about it and enact some change. Three of the four give us a pretty clear critique of traditional mathematics teaching, and the fourth questions *who* we are teaching and why. None of them particularly push us toward questioning more deeply *why* we are teaching mathematics. Some of them commit in one way or another to teaching mathematics for human capital, or those intangible qualities usable by businesses for their own profit generation. In selecting these books, I am demonstrating that the more mainstream critiques of mathematics education center mostly on reform mathematics teaching and sometimes on teaching for equity. This term "equity" is the signifier used by organizations like the U.S.'s National Council of Teachers of Mathematics that represents an increase in success in mathematics by students who are poor, nonwhite, and female. As we will see in future chapters, some of these operate under differing assumptions. In Moses's case, the goal is to prepare students for success *within* the current system in order to increase access to power and, perhaps, to change it. Such a conversation will be discussed in detail in Chapter 3 of this book on race and mathematics education.

As we move into the next section of this chapter, think about how these works have shaken up the typical view on mathematics instruction. It may seem as though there are two approaches to teaching mathematics—traditional and reform—and that a critical approach encourages reform more consistently. This is a good starting point for us to examine mathematics education in practice; we will look at two structures for lesson planning. The first resembles a traditional approach but will be greatly enhanced by the reform teaching style, and the second more fully embraces all that reform mathematics teaching offers. Throughout, we will question and be critical of all aspects of the lesson planning process in mathematics.

Critically Planning Mathematics Lessons

Acknowledging the practice of mathematics education is an important aspect to our work in critically examining it. We are often taught in general and mathematics-specific methods courses the dos and do nots of lesson and unit planning. In this section, we go through these specific elements of lesson plans as they are typically taught today but do so by examining each with a critical perspective. This aims to give you a framework within which you can design mathematics lessons that correspond to critical perspectives. This is an appropriate place to start in moving toward a critical practice; you will apply the methods learned here as you delve in much greater depth into the critical perspectives contained in future chapters regarding race, gender and social class.

The section will describe two structures for the lesson plan, one a traditional, direct-instruction approach that integrates reform mathematics teaching and the second, an experience-based, constructivist style to mathematics lesson planning. We begin with standards and how to think about these critically and then move through the components for each lesson plan structure. These include hooks, closures, direct-instruction modeling, experiences (or tasks), whole-class discussions, assessments, and the use of technology and other materials. Finally, we think critically about the inclusive strategies often promoted in mathematics teaching and learning, specifically how we can more effectively integrate language needs and students with learning disabilities.

Most lessons begin with a topic that you need to teach. Thinking uncritically about mathematics education means that you start with the standards and or textbook objectives that are given to you by the local education agency or school or other institution sponsoring your work. Critically teaching mathematics does not ask us to reject these standards entirely but instead to view them critically. This happens in a number of ways. First, recall our discussion of mathematics in the first chapter. Because mathematics is a social construction, we can extend the argument to mathematics standards. These are not objective, value-free assertions of "what everyone should know" but are instead social constructions. We will take a look at the politics of mathematics standards in the concluding chapter, but for now, our readings on philosophy of mathematics are enough to rattle the faith one might otherwise have in mathematics standards.

Another way to critically approach the standards is, when appropriate in your setting, you can rearrange or otherwise adapt standards to meet the specific learner needs and readiness in your classroom. Ultimately you must know more about your students than their previous standardized test score: You must talk with them, do mathematics with them, and ask them to do mathematics for you. Get inside their heads as much as you can. Emerging from the reform and mathematics education research movement, a wonderful tool at your disposal is the concept of the clinical interview, in which you observe one on one while a learner completes a problem-oriented mathematical task. Examples of these, including video, are easy to find using typical Internet search tools.

Yet another approach is to think about content outside of mathematics when designing your lessons. What educational content standards outside of mathematics might you want to teach in the lesson as well? Sometimes, a mathematical topic is taught much more effectively when given life with an application or real context. When teaching statistical topics, for example, this is almost always the case. You might want to pull in a standard about social justice from the U.S.'s National Council of Social Studies or perhaps an environmental science standard from Next Generation Science Standards. A mathematics education that is critical will not limit itself to mathematics content standards. To maintain your status as a mathematics educator, however, you will need to make sure that you can justify your lessons as maintaining a level of rigorous mathematical content. Using such standards strategically is at the heart of critical mathematics teaching, especially given today's prioritization of standards and assessment.

Once a lesson topic is chosen, the next step is to think critically about how you might want the class to function. Do you need to teach students a skill directly, or do you need them to discover it? This question works off the earlier section's discussion of reform and traditional teaching. With a topic at hand, I find it helpful to approach lesson design by next deciding whether a lesson will take on a direct-instruction (traditional-style) or experience-based (reform-style) approach. These are characterized with a rough outline of each format in Table 1.1. Although the former direct instruction does resemble the traditional style in many ways, we will discuss this within a reform approach to teaching mathematics. In this way, then, the two styles will both emphasize meaning making and an active role on the part of the learners. For example, one way that traditional mathematics teaching is *not* in the reform tradition is when students are blankly copying down the board work as the teacher is working out problems for the class or when students are filling in blanks on worksheets as the teacher goes over problems and ideas.

TABLE 2.1 Two mathematics lesson plan structures

Direct Instruction	*Experience Based*
1. Hook	1. Hook
2. Model	2. Experience
3. Guided practice	3. Whole-class discussion
4. Independent practice	4. Reinforcement
5. Closure	5. Closure

Keeping students active in their process of learning is a central tenet to teaching mathematics critically.

Both modes of lesson planning start with a hook, an opening to the instructional phase in which a stage is set for new learning. Drawing on understandings of how people learn, such as Piaget's notion of assimilation and accommodation, the hook should access the relevant prior knowledge a learner has for the lesson that is about to take place. This depends on the lesson's content and might take the form of a whole-class discussion about some context that will be related to mathematically or perhaps a mathematical task that students are fairly equipped to do already but need some refresher. Such an opening hook must adequately bring to the front those mathematical and other concepts/skills that will be required by students during the lesson. Equally important, the hook needs to engage the students and connect the teacher to the students. Establishing a community will not only settle everyone for the day's learning but pique curiosity as well. Question posing is particularly helpful for doing this.

Here are two hook examples for teaching a lesson in a precalculus or calculus class on limits. The first hook would ask students to simplify five rational expressions such as $\frac{x^2+4x+3}{x+1}$. Here, the students are required to factor trinomials and reduce the fraction by canceling common factors in the numerator and denominator. Students would have presumably learned this skill prior to the lesson, and such a review would bring this back to their minds. Once the limit is defined in the lesson, students will be able to get working right away on some limit problems. A second example of a hook for a limits lesson is to show a short clip from the movie *Mean Girls* in which popular actress Lindsay Lohan's character solves a limit problem in front of an audience (this clip is usually easy to find online). Assuming the students know who Lindsay Lohan is, this will establish community (imagine the discussions that might take place about popular culture) and, by introducing limits, get the class excited about the mathematics they are about to learn in the lesson.

Now compare these two hooks: The first one is rather dry, and for some students this will not motivate them for the lesson; the second might get students excited and establish community but will not bring to students' minds any relevant prior knowledge. The best lessons have hooks that fulfill both of these objectives. Sometimes it is not easy to find one hook that can cover both grounds, so you might provide two short hooks that complement each other in this way. Keep in mind your content objectives, both mathematical and otherwise, as you design your hooks. Make sure that your hooks are student friendly, establish a connection between students and between teacher and students, and are intelligently selected for their access of prior knowledge.

Moving through the parts of the traditional-style lesson next, the phase after the hook is the modeling phase. To be effective, such a direct-instruction-style lesson must incorporate active learning methods informed by cognitive research that tells us how people learn. In modeling, a teacher will explain the definitions, concepts, and skills directly to the students but will do so with an active student-centered approach. This takes on a variety of forms as well and depends on the learning goals. If the lesson is teaching a procedural skill, the notion of a think-aloud is highly appropriate. A think-aloud is when the teacher models the problem by noting every detail in their thought process as they complete it. They articulate out loud all the information they receive, all the choices they have, which choice they make, and why. Here is an example for thinking aloud how to graph an equation using a table of values, with actions and gestures the teacher takes put in italics. The equation to graph is $y = x^2 + 1$.

> I need to graph the equation $y = x^2 + 1$. *Points to equation.* Because the equation has only xs and ys and I need to graph it, this means I will need an x–y plane. *Draws a coordinate plane on the board.* Now, I need to think about what this will look like. *Pause.* I know that sometimes when I graph equations they turn out to be lines and other times they turn out to be curves. *Pause.* I'm not quite sure what this will look like because I haven't done enough of these to know a pattern yet. But I do know that graphing means plotting the points that "work" *(uses air quotes)* for the equation. I mean, I know that I need to find numbers that I can put in for the letters in the equation and then I can plot these on the graph and try to find a pattern. Let me see if I can try to find a pair that "works." So, how about putting a 1 in for x? *Takes a notecard with the number one on it and places it over the x in the equation.* So, putting a 1 in for x means that I need to square it and add 1. But which do I do first? Let me think. *Pause.* Order of operations tells me that exponents come before addition, so I square 1 first to get 1 and then add 1 to get 2. *Records this on the board.* This means that when x equals 1, y equals 2, and I think I need a way to record this better as I work with more numbers. *Pause.* Oh yes, I remember, I can use a table of values to record the pairs that "work" for the equation...

The think-aloud for the problem continues similarly. Envision a teacher performing this think aloud slowly, clearly, and with engagement with her audience so that the students are invested in the outcome. It is almost like a story being told. A word of caution: If a think-aloud is too overdone, students feel they are being talked down to. This is the fine line of the performance that is required to keep students engaged and actively learning.

There are several other methods to keep modeling active for learners. A teacher can question the students for what might come next by calling on students or stopping briefly for a think-pair-share. The latter is when the teacher poses a question to the class, requires students to think individually for a designated moment of time, next share their thoughts together for a designated amount of time, and finally pairs are called on to share their thinking. Think-pair-shares are highly effective for use in any part of a mathematics lesson and for either direct- or indirect-instruction lessons. A final way to keep students active in the modeling phase is the use of guided notes. However, these should prompt for students to thoughtfully respond on the page rather than fill in blanks or copy work from the board.

These possible methods for increasing active learning during the modeling phase are simply examples to move mathematics teaching in a critical direction that increases access to mathematical knowledge. As this book takes on the topics of race, class, and gender in future chapters, you will want to revisit the notion of increasing access when explaining mathematics to learners. For example, some believe that all mathematics instruction requires mathematically precise language. However, notice how in the think-aloud example, the language used did not use precise mathematical language with well-defined terms throughout. While mathematical accuracy and precision are important aspects of mathematical practice, a critical perspective on teaching mathematics recognizes that mathematical communication can take a variety of forms that induct learners *over time* into the practice of refined mathematical communication.

After the modeling section, a direct-instruction lesson requires students to apply the information they received immediately. This happens in two phases: guided and independent practice. There are several ways to structure this, and effective approaches emphasize quick feedback to learners and social engagement with other learners. In guided practice, a teacher can pose a problem to the class and have students work with partners to solve together. This can happen for several rounds before students work independently on a set of problems for individualized practice. Social learning is particularly important in both phases, as peers working together can negotiate meaning faster than in isolation. There are lots of ways to structure guided practice that will keep students active and engaged, and this depends greatly on the community that the teacher and students establish.

Both indirect and direct lesson plans end with a closure. Like the hook, closures contain several objectives, and you should strive for your lessons to close with as many of these aspects as possible. First, a closure needs to provide an opportunity for the student to walk away with a clear summary of what was learned during the lesson. This can be as simple as a teacher restating the main content objective one last time, but it is best to first have the class state this and the teacher refine it if necessary. As well, a closure can provide the teacher with a check for understanding. Finally, the closure connects back to the opening and relays what the next steps will be in how the material learned will be used either independently or in work together in the next lesson.

Moving to the makeup of an indirect, experience-based lesson, the hook and closure work much the same way. However, it is the insides of the lesson that require explanation, especially because this approach is new and often misunderstood. We now turn to examples from within the reform mathematics education canon because researchers in this vein developed this method carefully and with great detail. I will also point you to examples of social justice- and sustainability-oriented mathematics lessons that utilize the approach. As we will uncover, we must "move beyond not telling" as the main message to reform mathematics and instead carefully select tasks that match student readiness and that we can anticipate how to best facilitate our students' struggle on the tasks.

First, what do we mean by experience based? Here, the teacher relies on her knowledge of what the students know already and selects tasks that lead to students' uncovering of new knowledge. Using terminology developed in the research literature, the teacher "guides and facilitates, poses challenging questions, and helps students share knowledge" and the students "work in a group and learn actively" (Kuper and Kimani, 2013, p. 166). Many have misunderstood this to mean that teachers are "told not to tell," implying that teachers take an inactive role. However, this is entirely not the case; teachers must take an active role in selecting the task that will lead to learning, in anticipating how students will approach and work through the task and how they will probe and press students toward learning, and in orchestrating the subsequent whole-class discussion. For the latter, Chazin and Ball (1999) give an especially thorough account of the active role mathematics teachers must take to maintain the discussion's productivity and goal.

The central experience, or what many call the task, makes or breaks the experience-based mathematics lesson. Stein and Smith (1998) have codified the language of task selection into lower- and higher-level cognitive demand. The examples contained in their article provide descriptions of what kinds of tasks push toward the mathematics lesson that engenders significant, deep meaning making. As they describe it, a task is defined as a segment of classroom activity that is devoted to the development of a particular mathematical idea. A task can involve several related problems or extended work, up to an entire class period, on a single complex problem. Defined in this way, most tasks are from twenty to thirty minutes long" (p. 269). The higher cognitive demand categories are "procedures with connections" in which students make meaning as they recall and connect procedural mathematics and "doing mathematics," in which students uncover answers to previously unknown questions.

Take, for example, a middle school mathematics lesson in which students are to learn about the relationship among circumference, diameter, and pi. In a hook, students can be reminded of ratio and measurement, perhaps by being asked the question: "Compare these three rectangles' ratios of length to width. What do you notice?" Students will state that they have unequal ratios. This leads perfectly to the next prompt, serving as the central task of the lesson: Does the size of a circle change the ratio of its circumference to its diameter? Depending on student age and readiness with measurement, the teacher provides materials such as string and circles of various sizes. The students compare and contrast and debate the answer to the question until, lo and behold, the students determine that the ratio is about 3, or maybe just a little more than that. This lesson experience has simulated for students the process of "doing mathematics" as it was done over history and across several cultures in the conceptual development of what is now commonly termed pi.

Equally important to the selection of tasks is the orchestration of mathematical discussions that will unpack the experience of engaging with the task. This phase of the experience-based mathematics lesson has been studied significantly by researchers as well, such as in the article by Smith, Hughes, Engle and Stein (2009). Their research argues five practices to successful unpacking of experiences in the whole-class discussion:

1 Anticipating student responses to challenging mathematical tasks;
2 Monitoring students' work on and engagement with tasks;
3 Selecting particular students to present their mathematical work;
4 Sequencing the student responses that will be displayed in a specific order; and
5 Connecting different students' responses and connecting responses to key mathematical ideas (p. 550).

In teaching with a more critical perspective, it is important to remember the ways that social identities impact student participation in whole-class discussion. For example, we will look in detail in Chapter 4 at some mathematics education literature suggesting that such whole-class discussion models can unintentionally position working-class students as unsuccessful. As with all suggestions made by both reform and traditional teaching, knowing your students and their readiness will impact the structures you provide for learning. The fourth component of an experience-based lesson is reinforcement. After a whole-class discussion in which students unpack their understandings from the task, the teacher must provide further prompts, either mathematical tasks or perhaps a journal entry, that record and further solidify what was learned.

The final point regarding such experience-based mathematics lessons is not from scholarship of mainstream mathematics education but instead from broader critical perspectives. With their high-cognitive-demand tasks and whole-class discussion, task-oriented lessons need not only be oriented exclusively toward mathematics. In teaching mathematics with a critical perspective, choosing contexts by which the mathematics can relate can open opportunities for interdisciplinary work in mathematics teaching and learning. Critical work in education would suggest that we look to opportunities in social justice and sustainability issues. For examples of mathematics lessons that orient toward a reform style of mathematics teaching *as well as* critical contexts in which to apply and learn mathematics, take a look at the two books *Rethinking Mathematics: Teaching Social Justice by the Numbers* (2013) and *Teaching Secondary Mathematics as if the Planet Mattered* (2013). These books offer countless mathematics lesson plans across grade levels that require active students who will engage deeply in controversial issues.

Teaching for Inclusion: Linguistically Responsive Mathematics Teaching and Universally Designing Mathematics Lessons

Many times a lesson plan format contains a requirement to differentiate or accommodate for students, usually English language learners and students with learning disabilities. These are strong considerations that move toward teaching mathematics with a critical perspective, and to consider them in depth, we look at the more critical orientations for teaching for inclusion. We will look at mainstream resources for mathematics teachers and complement these with the most critical perspectives on language and learning needs that have yet to be applied to mathematics education.

Led by Bill Jasper, the Mathematics for English Language Learners Group in Texas has developed the *Teacher's Guide to Teaching Mathematics for English Language Learners* (2005). This resource is framed by a fairly critical viewpoint that aims to increase success in mathematics by what it terms English language learner (ELL) students. As we discuss their work, we will continue to use the term ELL. However, in a bit, we will look toward more forward-thinking terms like "emergent bilinguals" to denote these language-minority students. The authors frame their guide within an understanding of the multiplicity of challenges that ELL students face, including language, socioeconomic, and cultural differences and differing backgrounds in schooling in addition to varied prior knowledge in mathematics. The guide provides several important and practical suggestions for mathematics teachers, such as acknowledging and knowing the impact of certain cultural differences (this guide focuses on Spanish-speaking ELLs). As well, mathematics teachers must recognize important mathematical content differences, such as writing numbers differently and different setups for the subtraction and division algorithm. The guide suggests mathematics teachers must "value students' background, encourage family involvement and use appropriate phrases" (Jasper, Huber, and Mathematics for English Learner Group, 2005, pp. 8–9), such as greetings and simple instructions, in the students' first language.

Specific classroom practices from the guide include establishing a positive classroom environment, speaking slowly and clearly, careful selection and adaptation of materials for your students, and necessary assignment and assessment modifications. The guide encourages reform mathematics teaching when appropriate, encouraging a conceptual understanding because

> Timed drills or repetitive problems are rarely effective for ELLs and seldom help them retain mathematical knowledge for very long. Periodically during the lesson or at the end of the lesson, review important concepts and vocabulary, and connect these to the objectives for the lesson.
>
> (*p. 14*)

The guide also suggests the use of grouping and collaborative learning as an effective strategy for ELL students and other important tools, like the use of graphic organizers.

The guide provides a solid introduction to the research in mathematics teaching for ELLs and a resulting list of best practices. However, there are more advanced, critical conceptions of English language learners that mathematics teachers with a critical perspective should acknowledge and understand. Bilingual education expert Ofelia García denotes ELLs as "emergent bilinguals" rather than ELLs because the ELL signifier implies a move toward replacing a native language with English. She describes the linguistic experience of bilinguals and emergent bilinguals alike as translanguaging:

> Translanguaging is the act performed by bilinguals of accessing different linguistic features or various modes of what are described as autonomous languages, in order to maximize communicative potential. It is an approach to bilingualism that is centered, not on languages as has often been the case, but on the practices of bilinguals that are readily observable in order to make sense of their multilingual worlds.
>
> *(García, 2009, p. 140)*

Such a view of emergent bilinguals or language-minority students in our classrooms positions them as eager communicators drawing on multiple strands rather than as students who *need to have* English skills.

In applying translanguaging to teaching and learning, García describes the pedagogical implications of translanguaging as having two principles: social justice and social practice:

> The social justice principle values the strength of bilingual students and communities, and builds on their language practices. It enables the creation of learning contexts that are not threatening to the students' identities, but that builds multiplicities of language uses and linguistic identities, while maintaining academic rigor and upholding high expectations. Another important element of this principle has to do with advocating for the linguistic human rights of students and for assessment that includes the languaging of bilingual students. The social practice principle places learning as a result of collaborative social practices in which students try out ideas and actions, and thus socially construct their learning. Learning is seen as occurring through doing. Translanguaging among students, especially in linguistically heterogeneous collaborative groups, becomes the way in which students try out their ideas and actions and thus, learn and develop literacy practices.
>
> *(p. 153)*

These principles apply to mathematics classrooms and are consistent with reform mathematics' emphasis on collaborative learning and negotiating meaning. With these in mind, mathematics teachers shift their thinking from language-minority students as having deficits toward students with advantage.

Fortunately, the framework of translanguaging has been applied directly to mathematics education in the form of a sample unit plan. In their *Translanguaging in Curriculum and Instruction: A CUNY-NYSIEB Guide for Educators* (2014), Sarah Hesson, Kate Seltzer, and Heather H. Woodley provide a unit plan for middle school mathematics titled "A Task of Monumental Proportions." The unit focuses on the study and design of monuments with mathematical content areas of proportion, measurement, and geometry. Throughout the unit, the authors provide "translanguaging how-tos" built upon the theory and principles that García outlined by translanguaging and its pedagogical implications. For example, "Students are multilingual tour guides for each other on a field trip, or use a multilingual audio-tour" and "Provide home language math texts and word problems" (Hesson, Seltzer, & Woodley, 2014, p. 79). This unit and its translanguaging how-tos serve as a model for routine applications we can make in all mathematics units. Rather than "accommodating" for language-minority students, these practices can be embedded in our lesson plans consistently.

Similar to the development of approaches to teaching mathematics to language-minority students, mathematics education has advanced significantly in "accommodating" students with learning disabilities, and we can complement this work with more advanced and critical theory on disability. First, let's review the typical approaches to students with learning disabilities, with the currently-in-favor "response to intervention" model. A teacher guide that uses this model was put out by the U.S. Institute of Education Sciences: *Assisting Students Struggling with Mathematics: Response to Intervention (RtI) for Elementary and Middle School Students* (2009) provides how-to assistance in the early detection of and accommodation for learning disabilities as they relate to mathematics instruction. Situated within the efforts of the 2004 reauthorization of the Individuals with Disabilities Act, the guide's goal is to:

> provide suggestions for assessing students' mathematical abilities and implementing mathematics interventions within an RtI framework, in a way that reflects the best evidence on effective practices in mathematics interventions. RtI begins with high-quality instruction and universal screening for all students. Whereas high-quality instruction seeks to prevent mathematics difficulties, screening allows for early detection of difficulties if they emerge. Intensive interventions are then provided to support students in need of assistance with mathematics learning. Student responses to intervention are measured to determine whether they have made adequate progress and (1) no longer need intervention, (2) continue to need some intervention, or (3) need more intensive intervention.
>
> *(Gersten et al, 2009, p. 4)*

An example of the recommendations contained in this report is the emphasis on visual representations to augment understanding of mathematical ideas. This is consistent with reform mathematics pedagogy, in which multiple modes of teaching concepts are encouraged for all learners.

Although the intended efforts of RtI support students with disabilities, unintended consequences can result. Ferri (2010) critiques RtI as a tactic that reinforces exclusionary practices in school systems. She claims that RtI is the most recent iteration of special education reforms that objectify students with special needs:

> Once deemed eligible for special education, students are assumed to be 'fundamentally different' from their non-disabled peers. Disability labels, therefore, function as a discursively produced system of social othering that creates divisions between students who are considered normal and regular and those who are seen as deficient and disordered.
>
> *(p. 1)*

Unfortunately, the RtI mathematics guide from the Institute of Education Sciences commits these faults, and we need to look instead to efforts within mathematics education that are informed by what has been termed disability studies.

Several scholars work in applying disability studies to education, including Beth Kerri, Susan Gabel, and David Connor. As Gabel and Connor (2009) put it, disability studies education (DSE)

> recognizes [disability] as a natural form of human variation—one difference among many human differences—and better understood as the results of an interplay between the individual and society and through analyzing social, political, cultural, and historical frameworks. Most importantly, DSE is critical of beliefs and practices that produce inequalities in the social conditions of schooling.
>
> *(p. 379)*

DSE views disability as a social construction, calling into question how people labeled disabled are perceived as misfortunate, less than, and incapable. In this view, the not-normal disability label can be considered an oppressive activity similar to not white, not male, and so forth.

DSE contains several pedagogical implications, and our goal is to relate these to the mathematics classroom. First, curricular content can be more inclusive of people living with disabilities. "Whether or not district policy requires the use of particular textbooks, teachers usually have the flexibility to select materials of their own choosing while still adhering to state standards" (Gabel & Connor, 2008, p. 389). For example, a mathematics teacher can tell the stories of mathematicians with disabilities (a simple Internet search can reveal several biographies), doing so without objectifying those who live with disabilities. Gabel and Connor also suggest content that is critical of media descriptions of people living with disabilities, and we can imagine statistical projects that numerate these instances or survey popular opinion and aim to reeducate people about disability.

The second tenet to DSE's pedagogical implications involves access to curriculum and embraces a universal design for learning approach. Here, "the universal design literature echoes many of the principles of differentiated instruction but universal design deemphasizes documentation and labelling . . . and emphasizes the creation of environments accessible to the greatest number of people possible without regard to disability status" (p. 389). As we lesson plan for mathematics, we should have keen awareness of our students' needs and readiness. To be sure, universal design for mathematics does not suggest that specific diagnoses for individuals are to be ignored. We must consider all knowledge of our students that has been made available to us, including how they learn best. Using this knowledge, we plan active lessons and activities that take this into account without entrenching students in fixed groups or categories and entirely without labels. There are several active projects in designing universal mathematics lessons and units, such as the universal design for learning (UDL) project.

In the preceding sections, we have reviewed several of the typical components of lesson planning for mathematics instruction, doing so with a critical perspective. This began with thinking critically about how to use content standards, including the idea that we can integrate social justice and sustainability issues in our mathematics instruction. We next went through the components of two types of lesson strategies: direct and indirect. I encourage mathematics educators to consider both of these, albeit doing so critically with a conception that mathematics is a social activity, implying that people learn mathematics socially and through activity. The importance of hooks and closures was stressed so as to make sure mathematical learning is connected to other lessons and experiences in real life. Finally, we examined critically how we can provide inclusive instruction for emergent bilingual learners and learners with disabilities.

These latter efforts in particular lead us directly into the next three chapters of the book. Typically, language-minority and special-needs students are positioned as inferior and with deficits. However, a critical orientation promotes an entirely different orientation, in which educators understand these labels as social identities embedded in political, historical, cultural, and economic circumstances. Critical educators work to resist such labels of inferiority and reframe our work so as to provide inclusive education that dignifies all life and provides universal access to content. As we shall see in the upcoming chapters, great efforts in mathematics education have taken up the social identities of race, social class, and gender. In each chapter, we take on critical orientations in which these identities are understood as social constructions that aim to objectify and make inferior. Similarly, our goal is to design universal mathematics instruction that honors and provides access to, for example, the Black, the working class, and girls and women.

Activities and Prompts for Your Consideration

1 Discuss the books outlined in the first half of the chapter. Given the description contained here, which do you agree with and which do you disagree with? Select one to read a few chapters from and discuss with others whether your original opinion holds true.
2 Would you say that your mathematics teachers in the past have favored traditional or reform teaching methods? What style do you seem more comfortable with, and which seems harder?
3 Pick a mathematical standard for a grade level that you might be asked to teach. First, design a lesson that uses the experience-based format. Next, choose a content objective focused on social justice or sustainability and integrate it into your lesson plan. Finally, add one "translanguaging how-to" and one "universal design" feature.

References

Boaler, J. (2009). *What's math got to do with it? How parents and teachers can help children love their least favorite subject.* New York: Penguin.

Chazin, D. & Ball, D. (1999). Beyond being told not to tell. *For the Learning of Mathematics 19* (2): 2–10.

Coles, A., Barwell, R., Cotton, T., Winter, J. & Brown, L. (2013). *Teaching secondary mathematics as if the planet mattered.* New York: Routledge.

Ferri, B. (2010). Undermining inclusion? A critical reading of response to intervention. *International Journal of Inclusive Education 16* (8): 1–18.

Gabel, S. L. & Connor, D. (2009). Theorizing disability: Implications and applications for social justice education. In *Handbook of Social Justice in Education*, edited by William Ayers, Therese Quinn & David Stovall. New York: Routledge, pp. 377–399.

García, O. (2009). Education, multilingualism, and translanguaging in the 21st century. In *Multilingual Education for Social Justice: Globalising the local*, edited by Ajit Mohanty, Minati Panda, Robert Phillipson & Tove Skutnabb-Kangas. New Delhi: Orient Blackswan (former Orient Longman), pp. 128–145.

Gersten, R., Beckmann, S., Clarke, B., Foegen, A., Marsh, L., Star, J. R. & Witzel, B. (2009). *Assisting students struggling with mathematics: Response to Intervention (RtI) for elementary and middle schools* (NCEE 2009–4060). Washington, DC: National Center for Education Evaluation and Regional Assistance, Institute of Education Sciences, U.S. Department of Education. Available at http://ies.ed.gov/ncee/wwc/publications/practiceguides/.

Gutstein, E. & Peterson, B. (2013). *Rethinking mathematics: Teaching social justice by the numbers.* Milwaukee, WI: Rethinking Schools.

Hesson, S., Seltzer, K. & Woodley, H. H. (2014). *Translanguaging in curriculum and instruction: A CUNY-NYSIEB guide for educators.* New York: CUNY-NYSIEB. Available at www.cuny-nysieb.org.

Jasper, B., Huber, J. & Mathematics for English Learner Group. (2005). *Teachers guide to teaching mathematics for English language learners.* Available at tsusemll.org.

Kuper, E. & Kimani, P. (2013). Responding to students' work on a rich task. *Mathematics Teaching in the Middle School 19* (3): 164–171.

Lockhart, P. (2009). *Mathematician's lament: How school cheats us out of our most fascinating and imaginative art form*. New York: Bellevue Literary Press.

Ma, L. (2010). *Knowing and teaching elementary mathematics: Teachers' understanding of fundamental mathematics in China and the United States.* New York: Routledge.

Moses, R. P. & Cobb, C. E. (2001). *Radical equations: Civil rights from Mississippi to the algebra project*. Boston, MA: Beacon.

Smith, M. S., Hughes, E. K., Engle, R. A. & Stein, M. K. (2009). Orchestrating discussions. *Mathematics Teaching in the Middle School 14* (9): 549–556.

Stein, M. K. & Smith, M. S. (1998). Mathematical tasks as a framework for reflection: From research to practice. *Mathematics Teaching in the Middle School 3* (4): 268–275.

Wolfmeyer, M. (2014). *Math education for America? Policy networks, big business and pedagogy wars.* New York: Routledge.

3

A WHITE INSTITUTIONAL SPACE

Race and Mathematics Education

Since the summer of 2013, the United States' Black Lives Matter movement has been calling attention to and protesting the routine murder of black civilians through police brutality, as well as other racial injustices, such as the school-to-prison pipeline and disproportionate incarceration rates of African Americans. Their work continues the long struggle against white supremacy, drawing from the variety of groups and individuals in the civil rights movement from Dr. Martin Luther King Jr. to the Black Panthers. Simultaneously, Black mathematics educators like Danny Martin have been asserting that mathematics education, and even specifically groups like the National Council of Teachers of Mathematics (NCTM), are a racialized project conceived of by whites and in their interest and one that only superficially takes efforts toward addressing racial injustice as experienced through mathematics teaching and learning. This chapter aims to present to you the parallels in these pronouncements, to question how mathematics education perpetuates white supremacy and in what ways we can teach critically to interrupt this.

In the next three chapters on race, class, and gender, we begin with a broader discussion of related concepts before moving on to specific details and practices for what we know about teaching mathematics with a critical perspective. In this case, we first orient to the notion that race is a social construction and review the historical development of white supremacy. Next, we look at a sampling of concepts related to race and education in general. This motivates questions relevant to mathematics education, and we next take an in-depth look at the variety of writings related to mathematics education and race. Finally, some examples are provided to motivate your practice in teaching mathematics with a critical *race* perspective. Each of these sections is meant to introduce you to major concepts, and I encourage you to explore these more deeply, possibly by reading the suggested texts, engaging in the activities provided at the end, or developing your own inquiries and discussions.

White Supremacy and the Social Construction of Race

In beginning our efforts to teach mathematics with a critical race perspective, we must first understand the critical perspective on race, racism, and white supremacy. The notion that race is a social construction is the typical place to start. A casual experiencer of the modern world will no doubt have some basic understanding of or experiences with the notion of race. In the United States, the government is fond of categorizing people according to racial groups: The 2000 census reported the country's population according to "White, African American, American Indian/Alaska Native, Asian, Hawaiian Native & Pacific Islander" and various combinations of the list. At the very least, we experience race in some sense by checking off these boxes.

However, after this, experiences with race and racism differentiate rather starkly. Broadly speaking, some people (typically nonwhite) consciously experience racism every day, and others are subconsciously a part of racism but can decide to consciously engage with the concept of race or not. Take the latter, and I am referring to the white population. Whites have racial experiences ranging from the following: those who choose to have a critical, keen awareness of race and racism to those with an attitude/experience of colorblindness to those acting along the lines of (and sometimes admitting to) white supremacist ideology. In what follows, I introduce the contributions of three scholars, the first two describing how the construction of white supremacy came to be and the last contributing in words the social experience of being nonwhite in a racist society. Although the term "white supremacy" might conjure up the image of neo-Nazis protesting against, say, the Black Lives Matter movement, we are using it here to specifically denote how race as a social construct functions as a sociopolitical ordering of relations between groups of people. The phrase accurately describes the hierarchy contained with this ordering, with whites on top.

Historian George Fredrickson, author of *White Supremacy: A Comparative Study in American and South African History* (1981), prefers to use the term "white supremacy" because the word "racism" has the disadvantage of having

> been used so frequently as an epithet. No one, at least in our time, will admit to being a racist. The phrase white supremacy, on the other hand, is relatively neutral; both defenders and opponents of a fixed racial hierarchy have been willing to invoke it.
>
> *(p. xxii)*

In a sense, the word "racism" invokes knee-jerk responses in some without clear understanding of what we are talking about; it also opens the door for discussions of reverse racism, a circumstance that has not had the impact on whites in any sustained way that it has had on Blacks. "White supremacy" reminds us what we are talking about: a habit of mind in which whites on the whole are characterized as superior to nonwhites. As Fredrickson says it,

> white supremacy refers to the attitudes, ideologies, and policies associated with the rise of blatant forms of white or European dominance over 'non-white' populations. In other words, it involves making invidious distinctions of a socially crucial kind that are based primarily, if not exclusively, on physical characteristics or ancestry.
>
> *(p. xi)*

Fredrickson's book studies how this habit of mind historically *came to be*, making the claim that the developmental history of white supremacy in the United States was entangled with economic opportunity and social-political development. On the one hand, white supremacy came to be because European settlers in the United States South needed justification, a reason, to enslave people of African descent, and continued long after the emancipation of slaves in 1865. Fredrickson is careful to note, however, that economics did not entirely drive racial divides. Instead, the rise of ethnic consciousness and economic advantage fed off each other in the development of racial and class distinctions over time.

Another scholar, scientist Stephen Jay Gould, provides a complement to Fredrickson's study on the historical development of white supremacy. *The Mismeasure of Man* (1996) tells the chronology of the mathematicians and scientists who attempted to validate white supremacy for white supremacy's sociopolitical and economic needs. Essentially, it was a quest in biological determinism or to prove that "shared behavioral norms, and the social and economic differences between human groups—primarily races, classes, and sexes—arise from inherited, inborn distinctions and that society" (p. 52). Over the years, efforts to document differences in brain size and certain styles of psychological testing all claimed to scientifically document white supremacy. One by one, Gould picks apart these studies for their flaws, from 19th-century studies of brain size to the 1990's *Bell Curve* (Herrnstein & Murray) argument relying on IQ testing.

To understand race as a social construct is to take a hard look at its development, as these two have done. By saying that "Race is a social construct," it means that the racial categories have no scientific viability. As documented in the histories mentioned, the superiority of the white race was the result of historical developments in geographies, politics, and economics that necessitated claims of superiority with respect to ancestry and skin color.

Race is a social construct: There is no gene or other aspect of biology that determines one's race. Racial categories are the result of historical developments and associated shifts in place, power, and economics. Race and racism impact people significantly.

Saying that race is a social construct is not the same thing as saying "Race does not exist." Because it is a social construct, it has sustained effects on people and how they interact. We are all affected by race and racism, and it is difficult to escape, especially by simply stating, "I am not a racist." It requires checking the habits of mind and actions resulting from white supremacist logics. For example, take a mathematics teacher who teaches mathematical reasoning to his white students and mathematical mechanics to his Black students. This teacher might be acting on an unchecked, subconscious attitude of white supremacy, causing him to expect more from his white students and less from his Black students.

Reviewing the development of white supremacy sheds light on the habits of mind for whites in a racist society. Teaching mathematics with a critical perspective requires white teachers to ask themselves: What does it mean to be white in a racist society? On the other hand, both whites and nonwhites need to understand the habits of mind for nonwhites who live within white-supremacist power structures. For these we turn to the classic text by W.E.B. DuBois.

Written in 1903, *The Souls of Black Folk* describes nonwhite, specifically, Black consciousness in the United States. He opens the first chapter of the book by describing what it feels like "to be a problem" and describes the Black psyche as follows:

> After the Egyptian and Indian, the Greek and Roman, the Teuton and Mongolian, the Negro is a sort of seventh son, born with a veil, and gifted with second-sight in this American world,—a world which yields him no true self-consciousness, but only lets him see himself through the revelation of the other world. It is a peculiar sensation, this double-consciousness, this sense of always looking at one's self through the eyes of others, of measuring one's soul by the tape of a world that looks on in amused contempt and pity. One ever feels his twoness,—an American, a Negro; two souls, two thoughts, two unreconciled strivings; two warring ideals in one dark body, whose dogged strength alone keeps it from being torn asunder.
>
> *(DuBois, 1994, p. 11)*

DuBois's description of twoness points to the strong identity that race places on nonwhites in society. A white teacher who states, "When I see people, I do not see color" simply showcases that white folks cannot know and feel what it is like to have a racial identity. Quite the contrary, a nonwhite person in a racist society is always conscious of their racial identity.

The goals of this section included reviewing the concept of race as a social construct so as to consider it as a habit of mind taught to us that shapes our social experiences. We have looked at some of its origins and thought about how it affects individuals and society. Next we will apply the notion of "race as social construct" to education more generally. At this point, continue to think about whether mathematics education plays a role in white supremacy. Most would argue that mathematics, as a so-called neutral subject, endures less impact from racism than the other subjects taught in schools. As we shall see in the final section of this chapter, there is little support for this claim.

Race and Education: Critical Race Theory and Culturally Relevant Pedagogy

This section introduces you to some practices and theories in education that take up the notion of white supremacy and racial constructs directly. We begin with an understanding of critical race theory and move to examples of practice from culturally responsive pedagogy. We also review debates within the conversation on race and education, such as the interplay that race has with capitalism and whether education should address the two together or separately.

Critical race theory emerged as a field within legal studies and has since been taken up by a great many critical education scholars and practitioners. Gloria Ladson-Billings and William Tate provide an introduction to its application to schools with the article "Towards a Critical Race Theory of Education" (1995). Keeping in the critical race theory tradition, the following three tenets are put forth:

> 1) Race continues to be a significant factor in determining inequity in the United States. 2) US society is based on property rights. 3) The intersection of race and property creates an analytic tool through which we can understand social (and, consequently, school) inequity."
>
> *(p. 48)*

Many educational theorists and practitioners have applied critical race theory to critique mainstream education and to imagine antiracist schooling experiences. Applying critical race theory to education requires questioning the foundation of the liberal order via an analysis of property rights and its relationship to race, education, and inequity.

In other words, inequities in property holdings coexist with inequities in power. Connecting this to education, Ladson-Billings and Tate (1995) suggest property's relevance to unequal schooling opportunities. For example, where a student goes to school is determined by her family's residence, and this affects the type of schools she attends. Although official desegregation occurred in the 1960s, de facto segregation occurs via "white flight" as a particular school district's community changes racial demographics. Other processes and structures related to property and real estate have also led to segregated schools, such as redlining, the practice through which particular neighborhoods exclude nonwhites by preventing mortgages for the purchase of homes. Such analysis runs counter to common assumptions about schooling; public education is assumed to level the playing field by providing all with equal opportunity to life's chances.

Critical race theory: Applied to education, critical race theory questions the interrelation among race, power, and inequity and begins by looking at unequal access to quality schools.

It is not enough to suggest that the suburban–urban divide, tracking within diverse schools, and the like, occurred as the result of economic divisions within society. Ladson-Billings and Tate (1995) write:

> If racism were merely isolated, unrelated, individual acts, we would expect to see at least a few examples of educational excellence and equity together in the nation's public schools. Instead, those places where African Americans do experience educational success tend to be outside the public schools. Some might argue that poor children, regardless of race, do worse in school, and that the high rate of poverty among African Americans contributes to their dismal school performance; however, we argue that the cause of their poverty in conjunction with the condition of their schools and schooling is institutional and structural racism.
>
> *(p. 55)*

Ladson-Billings and Tate carefully argue against any notion that poverty in its entirety causes the poor educational outcomes for nonwhites in this country.

Zeus Leonardo, another expert on race and education, suggests it is near impossible to say these disparities are entirely economically based. He also cannot suggest that it is entirely race based. The two have coevolved and continue to coexist. A capitalist society requires inequities; racial injustice feeds itself and also perpetuates economic disparities. He writes,

> Often, when Marxist orthodoxy takes up race, it reduces race relations to the status of a reflex within class dynamics... On the other hand, when race analysis takes up class issues, it sometimes accomplishes this by reifying race as something primordial or fixed, rather than social and historical.
>
> *(Leonardo, 2009, p. 45)*

For example, Leonardo takes issue with the fact that some educational theorists privilege the economic circumstances of Blacks over their status in a racialized society when deciphering the roots of disparities in school performance. Leonardo argues a "double jeopardy," in which working-class students of color "confront specific interlocking conditions of class exploitation and racial stratification" (2009, p. 48). Critical race theorists and others looking critically at race and education typically do see such intersections between race and class, but I agree with the critical race theorists who suggest that it's important that we not subsume the one (race) under the other (class). Saying that racial disparities in education are simply economic circumstances ignores entirely that race is constructed as a social hierarchy, as discussed in the first section of this chapter.

Moving from analysis and toward pedagogies that interrupt racial injustice, those within the critical race education camp take on a variety of stances. These run the spectrum from those arguing for individual success in the system to those calling for revolutions of society. We will briefly review two scholars who might represent this spectrum. On the one hand, Lisa Delpit writes of the ways that progressive education privileges white students; on the other, bell hooks suggests a pedagogy for change, akin to Paulo Freire's critical pedagogy. A third opportunity, culturally relevant pedagogy, comes again from Gloria Ladson-Billings.

Lisa Delpit (1995) cautions educators against enacting any practice that perpetuates racial injustice. In particular, she is concerned with the ways that so-called progressive education sets up a system in which white students succeed at greater rates than Black students. As an expert in literacy education for African American youth, she specifically argues for teaching students the "language of power," in her case standard American English. Limiting African Americans from learning this and other aspects of the "culture of power" is a racist project because these students fail to gain access to power. Without knowing codes of conduct (behavior and language, for example) for success, students will have little access to powerful positions in economic and political life. The sense here is that Delpit suggests antiracist education for a leveling of the playing field. We must keep this in mind when thinking about the progressive reform mathematics teaching. In what ways might we implement this pedagogy and, unintentionally, reinforce racial injustice through this practice?

On the other hand, bell hooks (1994) suggests a Black feminist pedagogy in the spirit of revolution. hooks is situated more clearly on the side of critical pedagogy because she looks deeply at societal structures and patterns in the hopes of change rather than in the reproduction of class structures, winners and losers. She was a student of Paulo Freire and came to write on the ways education can practice freedom. Focusing on domination as the root cause of oppression came to be a central tenet in such teaching praxis. In her book *Teaching to Transgress*, hooks articulates a distinction between education that practices freedom and education that reinforces domination. She would surely embrace several of the points made by Lisa Delpit, but it seems possible that Delpit's focus on gaining power is in tension with an education that practices freedom. These two arguments are offered to provide a spectrum of the approaches considered within critical education that attends to race and ethnicity.

Culturally relevant pedagogy: As a method of teaching nonwhite students, culturally relevant pedagogy has three facets: academic excellence, cultural competence in one's own culture, and a critical consciousness in coming to know the world and one's place in it.

A third, more specific and, we might say, practiced option is that of Ladson-Billings's culturally relevant pedagogy. Somewhat along the lines of Delpit's accessing the language of power as well as hooks's call to change, this three-faceted approach to teaching nonwhite students aims to increase student success while maintaining dignity and respect for one's culture. The three facets are (1) academic success, (2) cultural competence, and (3) critical consciousness, and I include her introductions of each from her initial article on culturally relevant pedagogy, "But That's Just Good Teaching: The Case for Culturally Relevant Pedagogy" (1995). Academic success:

> Despite the current social inequities and hostile classroom environments, students must develop their academic skills. The way those skills are developed may vary, but all students need literacy, numeracy, technological, social, and political skills in order to be active participants in a democracy.
>
> *(p. 160)*

Here she agrees with Delpit that no educational measures provided to nonwhite students should inadequately prepare them for economic circumstances as they are today. Second, cultural competence:

> Culturally relevant teaching requires that students maintain some cultural integrity as well as academic excellence. In their widely cited article, Fordham and Ogbu point to a phenomenon called 'acting White,' where African American students fear being ostracized by their peers for demonstrating interest in and succeeding in academic and other school related tasks.
>
> *(pp. 160–161)*

Culturally relevant pedagogy provides education for nonwhite children in which conscious attempts are made to embrace and not reject one's own culture. Finally, critical consciousness:

> Culturally relevant teaching does not imply that it is enough for students to choose academic excellence and remain culturally grounded if those skills and abilities represent only an individual achievement. Beyond those

> individual characteristics of academic achievement and cultural competence, students must develop a broader sociopolitical consciousness that allows them to critique the cultural norms, values, mores, and institutions that produce and maintain social inequities. If school is about preparing students for active citizenship, what better citizenship tool than the ability to critically analyze the society?
>
> *(p. 162)*

Along the lines of hooks's calls to change society, this facet of culturally relevant teaching directs teachers to expose children to injustice in the world.

As we move into our discussion on mathematics education and race, consider the possibilities that these three scholars have presented. First, how is mathematics education set up as a racialized space? Second, how do we teach to interrupt this space? Do we consider the role mathematics education plays in society and best prepare students of color to access this? Or do we reject mathematics education as the power elite conceives it and, along the lines of hooks's revolutionary spirit, teach an entirely different mathematics program?

What White Mathematics Education Is and How to Interrupt It

So far in this chapter, we have looked broadly at race and white supremacy in society and next applied this to education in general. Here, we take the next step in thinking about race and our efforts in teaching mathematics with a critical race perspective. We start with a look at the racialized nature of mathematics education policy and distinctions among those who discuss race in mathematics education. Next we look at several considerations and projects aiming to increase access and equity with respect to mathematics education. These contributions correspond directly to the theory and approaches laid out in the previous sections, including critical race theory and culturally relevant teaching.

Danny Martin, a professor of mathematics education, provides clear discussions about the racialization of mathematics education policy in the United States. His article "Race, Racial Projects and Mathematics Education" (2013) foregrounds the construct of race in understanding the historical development of math teaching and learning, at least as it is talked about at the top level. He argues how mathematics education can be characterized as "white institutionalized space" (p. 323). This means that white actors have dominated the definition and purpose of mathematics education over the course of its development. They have held and continue to hold math education's positions of power. Finally, mathematics education has long been portrayed as neutral, impartial, and unconnected to race and white supremacy. The last is perhaps the easiest to see firsthand. There has been a long tradition of actively stating that mathematics is not relevant to race, and yet race is so clearly related to student performance in it. And yet, how often do we say and hear claims of math as the "universal language," the objective, value-free knowledge? In this sense, we might say that mathematics education, as it has been conceived, should be more accurately called white mathematics education.

A compelling example of mathematics education as white institutional space is the tracked experiences that students endure in their schooling experiences. Students are placed according to their performance on standardized tests and teacher recommendations. The research article by Valerie Faulkner and her colleagues at North Carolina State titled "Race and Teacher Evaluations as Predictors of Algebra Placement" (2014) sheds insight into how the tracked experiences entangle with race and implicit biases. Essentially, their study reports, "Black students had reduced odds of being placed in algebra by the time they entered 8th grade even after controlling for performance in mathematics" (p. 288). The study calls our attention toward the impact of teacher recommendations for placement, and the researchers are careful in suggesting that such outcomes point to implicit bias in a white-supremacist society rather than conscious actions taken by individual teachers (although this certainly does happen).

No discussion of mathematics education and race would be complete without attention to the notion of the achievement gap. The gap represents the perceived distinctions in educational outcomes by white and nonwhite students, as reported by collecting and disseminating analysis on standardized tests. To be fair, it does also reference other social identities like students with disabilities and social class, but it is the racial categories that receive the most attention. Questions exist at the national as well as local levels. In my own district where I taught for 10 years in New Jersey, the superintendent consistently reminded us of the achievement gap *within the district*. In this case, students had the same materials, teachers, course offerings, and so on, and yet African American students would pass the standardized test in mathematics at a 60% rate and white students with a 90% rate. As a young mathematics educator who was teaching critically, I always had problems with the phrase "achievement gap" because I didn't agree that such measures reported a difference in achievement. Could not a student do well with little effort simply because of their privilege? How exactly is this a difference in achievement? With more nuance and detail, mathematics education researcher Rochelle Gutierrez critiques the phrase in her article "A 'Gap Gazing' Fetish in Mathematics Education?" (Gutierrez, 2008). She troubles the phrase in many ways, including how it highlights the perceptions of "better" and "worse" without any discussion of the bias that might be contained within the assessments. In this way, the attention the achievement gap receives actually perpetuates the notion of white supremacy. In place of this, Gutierrez focuses our attention on examples when education is working, when students of color are advancing mathematically.

Achievement gap: The conversation regarding the differentials in math test scores between white and nonwhite students. Stated to motivate our attention toward better education for all, but sometimes perpetuates myths about students of color, such as the inferiority of their minds or the defi cits in their home lives.

In this vein, another article by Rochelle Gutierrez draws our attention away from such deficit thinking and focuses on successful mathematical education for students of color. "Advancing African-American, Urban Youth in Mathematics" (2000) provides a case study of one mathematics department who consistently advanced urban students of color into higher-level mathematics courses. In her study, Gutierrez attributes this to "five characteristics of the department: a rigorous curriculum and the support to maneuver through it; active commitment to students; commitment to a collective enterprise; a resourceful and empowering chairperson, and standards-based instructional practices" (p. 63). She found the department to embrace the tenets of culturally relevant pedagogy as well, as discussed in the earlier section.

Complementing this, the article "Communities For and with Black Male Students," by C.C. Jett, David Stinson, and Brian Williams (2015), presents four strategies found to be consistent among successful teachers of African-American males. As they suggest, effective mathematics teachers for students of color are those who

> develop caring relationships that reach beyond the classroom, access and build on out-of-school experiences and community funds of knowledge during instruction, implement culturally relevant pedagogy throughout instruction, and disrupt school mathematics in particular and mathematics in general as a white institutional space.
>
> *(p. 286)*

Consistent application of Ladson-Billings's culturally relevant pedagogy is seen here, and by this we are encouraged to think deeply about what this looks like in the mathematics classroom.

A review article on the variety of research efforts made with respect to interrupting mathematics education as a white institutional space comes from the article "Rethinking Teaching and Learning Mathematics for Social Justice From a Critical Race Perspective" (2016) by Gregory Larnell, Erika Bullock and C. C. Jett. In the article, the authors position two definitive origins for the work of interrupting white mathematics education: Marilyn Frankenstein and Eric (Rico) Gutstein's work that applies Paulo Freire's pedagogy to mathematics and Bob

Moses's Algebra Project, aiming to increase the success in algebra by students of color (p. 20). The second is a book we looked at in Chapter 2, and the first will be discussed at greater length in the chapter on social class. Both aim to interrupt white mathematics education but in very different ways. The former, a Freirian critical pedagogy approach, redefines math education by using mathematics to read the world's injustice as well as using mathematics to rewrite the world. Consider how this approach resonates with bell hooks's more revolutionary spirit. On the other hand, "the Algebra Project regards the reality of college preparation as a central objective and mathematics becomes a tool or 'weapon in the struggle'" (Larnell, Bullock, & Jett, 2016, p. 22). We can view mathematics as an aspect of Delpit's "language of power," and to deny student's access to this is to perpetuate inequity across racial (and other) lines.

Larnell and colleagues do not present these as two opposing perspectives that we must debate on and decide which will lead to the greatest interruption of white mathematics education. Rather, these are two approaches that resonate with the broader conversations regarding race and education, such as the contributions of Lisa Delpit and bell hooks. Likely, both are equally important. Let's say that a high school math teacher in an urban school teaches juniors and seniors of color and aims to interrupt white mathematics education. He entirely replaces a white mathematics education program (and its learning objectives) with a course of study rich in the inquiry of injustice, perhaps through statistical investigations of local circumstances or the mathematical modeling of social relations. This no doubt interrupts white mathematics education and demonstrates to students the power of mathematical thinking. However, the teacher might not be preparing his students for college-level mathematics requirements. This circumstance, in which many students are prevented from obtaining a college degree because of their lack of proficiency in mathematics, is a problem that we can work to change but for the time being is firmly in place. Thus both providing greater access to white mathematics education, what we might call mathematical empowerment, as well as opening minds to mathematics beyond white mathematics education, what we might call revolutionary mathematics, are equally important aspects to the interruption of white mathematics education. Finally, the three tenets of culturally relevant teaching outlined in the previous section are clear objectives that we can apply as interruptions to white mathematics education.

In this chapter, we began by outlining the historical, cultural, and political elements of white supremacy. This revealed race as a socially constructed sociopolitical relation of power between groups, with an emphasis on the assumptive natures about inherited distinctions based on racial categories. In looking at the applications of such a critical conception to education, we reviewed the works of Zeus Leonardo, Lisa Delpit, and bell hooks, as well as Ladson-Billings's contributions of culturally relevant teaching and critical race theory. In connecting race with mathematics education, we have reviewed the arguments that mathematics education is a white institutional space in which racial injustice is only superficially addressed. Finally, several examples of teaching mathematics with a critical race perspective provide us with clear orientations and perspectives in moving toward teaching mathematics more inclusively and critically.

As a final note, I urge you to take up the cause of racial injustice as you see fit and in a manner appropriate to your identity. As an example, I'll share my own experience as a white individual. My path began with my own facing of race through study and listening to people. I do not assume that people of color are responsible for educating me about race. With regard to action, given my racial identity, I have found the concept of allyship to be crucially important in my efforts to fight racial injustice. This means several things, and you can take a look at several documents online to help clarify. Here are a few implications of what white allyship means to me: As a white individual, I cannot know the experience of not being white or know what must be fixed to solve the problem. I support movements like Black Lives Matter but do not lead in them. Instead, my work means that I educate white people about white supremacy and racial injustice. Finally, I aim to consistently reject white supremacist tendencies yet acknowledge the privilege that I receive by living in a white body.

Activities and Prompts for Your Consideration

1 The following is a semifictional case study of mathematics teaching. Read and discuss how this case does or does not interrupt white mathematics education. Does the case correspond to any of the following: culturally relevant teaching, revolutionary mathematics, or mathematical empowerment?

Mr. Jones is a white algebra teacher working in an urban/suburban school in the northeast United States. The student population is approximately 60% African American and 35% white. About 40 to 50% of the students qualify for free or reduced-price lunch.

In this district, students are tracked in classes based on test scores and teacher recommendation. This case focuses on Mr. Jones's lowest-level algebra 1 class. There are 18 students total: 15 African American, 3 white, 13 boys and 5 girls. In the last two weeks of May, the district curriculum requires this class to review number plots and learn about measures of central tendency (mean, median, mode) and spread (range).

After students finish their test on the previous unit, Mr. Jones begins this unit with the remaining 10 minutes of the class period. He holds a whole-class conversation about the students' preferences in popular music; it becomes clear that most students like hip-hop. Although some students may prefer R&B to hip-hop, all students in the class have great familiarity with the hip-hop genre. He then assigns students to bring in a favorite hip-hop song to share with the class the next day and requires that the songs are free of explicit lyrics.

In the next lesson, Mr. Jones begins with a think-pair-share: "What makes some hip-hop songs better than others?" At the opening activity's conclusion, Mr. Jones summarizes that the class thinks songs can be rated on the following: beat, flow (of lyrics), catchy hook, lyrics theme, and overall rating.

The students next use these categories to rate six of the songs that were brought to class that day. Students record their ratings on their sheets and put them on the board. Mr. Jones prompts students to think how they should display the data. After a brief whole-class discussion, the class decides that a number plot (which they learned earlier in the year) would work best. The day's lesson concludes with students recording data on number plots.

The next day, Mr. Jones asks students to discuss the data in groups. He prompts them for productive discussion: "What song did we seem to like the best? How can you tell that?" He also prompts students to think this through for each of the different categories (beat, flow, etc.). Students excitedly discuss and debate the answer to these prompts. Some have relevant prior knowledge and suggest finding the "average rating" for each song. Other students ask about how to do this and are taught by each other, with Mr. Jones facilitating.

In the ensuing whole-class discussion, a debate emerges. Roughly half the class thinks Song A is the class favorite because its average is highest. But some students, who happen to like Song C more, are arguing that Song C is the class favorite because the most common score it received was a 5/5 for the overall rating. Throughout this and previous conversations, Mr. Jones and the class also debate the themes of the six songs. Some of the students in the class openly disagree with a few songs' themes: emphases on "gang life," consumerism, and/or treatment of women. Mr. Jones facilitates discussions so that students' musical preferences are simultaneously honored and problematized.

At this point Mr. Jones interjects with some specific points to make. He congratulates the class for identifying the need to compute averages and for helping each other remember how to do this. He also highlights their discovery of another important feature in the data: looking for the value with the **most** responses. He directly teaches that these two **statistics** are referred to as **mean** and **mode**.

In a similar fashion, Mr. Jones prompts students to discover the concepts of **median** as well as **range**, a measure of spread. After a week and a half, Mr. Jones is satisfied that the students have learned the required concepts of the unit as well as practiced them on the hip-hop and other data sets. Because these objectives are finished within less time than allotted for the unit, Mr. Jones pushes the students to think more about measures of spread in the following final two days of instruction.

Mr. Jones says to the class: "You have noticed that for some songs we agreed quite a bit, but for others we disagreed. Which song did we agree the most on? How do you know?" In small groups, students use the range but are prompted by Mr. Jones to think about computing a "typical distance from the mean." Ultimately, students discover a method to average these distances. In a whole-class discussion, students share their methods, and Mr. Jones clarifies how to compute **absolute deviation**.

In the final instructional day, Mr. Jones reviews absolute deviation, both how to compute it and what it means. He shares with the students that this is material deemed too advanced for this class, and asks if they want more. Given their sustained enthusiasm, he next shares with students how to compute **standard deviation** for their hip-hop data. As they complete this, he reminds the students that standard deviation tells us the "average distance from the mean" in a similar way to absolute deviation.

At the conclusion of the unit, Mr. Jones is satisfied with his students' mastery of district curriculum as well as advanced material. In final discussions of the unit, students continue to reflect on their more nuanced understanding of hip-hop music. Students continue to debate themes in hip-hop as well as what makes a song sound good.

2 Because race is a social construct, we as individuals receive constant messages early on and throughout life that suggest how we should view the racial categories. One activity that is suggested to help us see this is the implicit association test. The point of this activity is not to place blame on individuals for their implicit associations but for them to understand these subconscious habits of mind so as to better know oneself and the tendencies that might occur without such introspection. Take the test at the link below and discuss with other mathematics educators the questions that follow:

Go to https://implicit.harvard.edu/implicit/takeatest.html and click through to find "Race IAT."

Questions

(a) Did your implicit association surprise you?

(b) Reflect on the times when you were implicitly taught white supremacist habits of mind, such as that whites are smarter or that nonwhites typically break the law. How are these learned by images and ideas from the media, actions our family members take, experiences in school?

(c) Anticipate how this relates to mathematics teaching and learning. How would a teacher's subconscious habits of mind affect learning outcomes?

3 Locate a mathematics teacher who is regarded for being successful with students of color. Perhaps the teacher: does not kick their students out of class or "call security" on them; incorporates racial justice issues into their

mathematics teaching; has a strong relationship with parents of color. Interview and/or observe the teacher, comparing their particular successes with the teachers in Gutierrez's (2000) and Jett, Stinson and Williams (2015).

4 Find a school with tracking for ability in mathematics (today this happens at all levels, including elementary, middle, and secondary). Observe a few of the tracked classes and/or interview a teacher or other school official about tracking. Collect data specifically for the possibilities that tracking falls along racial categories as well as the specific expectations required for each track. Typically, instruction in lower math tracks is less "time on task" and emphasizes skills rather than higher-order thinking. How does the mathematics tracking you see in the school relate to the concept of "white mathematics education" discussed in this chapter?

References

Delpit, L. (1995). *Other people's children: Cultural conflict in the classroom.* New York: New Press.

DuBois, W.E.B. (1994). *The souls of black folk.* Mineola, NY: Dover Thrift.

Faulkner, V., Stiff, L., Marshall, P., Nietfield, J., & Crossland, C. (2014). Race and teacher evaluations as predictors of algebra placement. *Journal for Research in Mathematics Education 45* (3): 288–311.

Fredrickson, G. (1981). *White supremacy: A comparative study in American and South African history.* New York: Oxford University Press.

Gould, S.J. (1996). *The mismeasure of man, Revised and expanded.* New York: Norton.

Gutierrez, R. (2000). Advancing African-American, urban youth in mathematics: Unpacking the success of one math department. *American Journal of Education 109* (1): 63–111.

Gutierrez, R. (2008). A 'gap-gazing' fetish in mathematics education? Problematizing research on the achievement gap. *Journal for Research in Mathematics Education 39* (4): 357–364.

hooks, b. (1994). *Teaching to transgress: Education as the practice of freedom.* New York: Routledge.

Jett, C., Stinson, D. & Williams, B. (2015). Communities for and with black male students. *The Mathematics Teacher 109* (4): 284–289.

Ladson-Billings, G. (1995). But that's just good teaching: The case for culturally relevant pedagogy. *Theory Into Practice 34* (3): 159–165.

Ladson-Billings, G. & Tate, W. (1995). Toward a critical race theory of education. *Teachers College Record 97* (1): 47–68.

Larnell, G., Bullock, E. & Jett, C. (2016). Rethinking teaching and learning mathematics for social justice from a critical race perspective. *Journal of Education 196* (1): 19–30.

Leonardo, Z. (2009). *Race, whiteness and education.* New York: Routledge.

Martin, D. (2013). Race, racial projects, and mathematics education. *Journal of Mathematics Education 44* (1): 316–333.

4

SOCIAL CLASS HIERARCHIES AND MATHEMATICS EDUCATION

To Reproduce or Interrupt?

How does today's socioeconomic order affect education? Should mathematics education allow for social mobility or reproduce class structure? How can teaching and learning mathematics interrupt the logics of capital? We will explore these questions throughout this chapter, looking critically at how traditional mathematics teaching limits social mobility at best and functions as a reproducer of social inequity at worst, as well as the instance in which reform mathematics falls short as well. As with the previous chapter on race, we open with broad social theory and move next to its application to education and finally to mathematics education. We begin this chapter with a broad discussion of the social construct of class, a structure that emerges anywhere capitalism is the dominating economic framework. This discussion, as with race, posits social class as a sociopolitical arrangement of social groups that serves to reproduce a hierarchy of social arrangements, with those in power remaining firmly at the top. Next we look at the intersection of class and education by reviewing scholars of education who prioritize the effects of class structure in the school system and vice versa. The schooling system, as it turns out, functions as a state structure to reproduce social-class hierarchy. These authors argue primarily that capitalism requires a reproduction of class hierarchies and uses the school system to do so. Finally, we look at critical work in mathematics education that focuses on social class in order to motivate the practical activities that close out the chapter. In this way, we will consider the means by which mathematics education is in service to social reproduction and actions we can take to critically work against this.

Critically Understanding Capitalism

In this section, we look to class as a social construct that emerges due to the logics inherent to capitalist economic order. This begins with general critiques of capitalism, from the classic text of Karl Marx to contemporary scholar David Harvey.

Overall, criticism rests squarely on the false pretense that capitalism is a fair competition with winners and losers. On the contrary, the logics of capital require a class hierarchy in order to maintain it, and the government helps with this role, in part by providing a public education that stratifies the population. As you read, think about what role education, and mathematics education, might play in maintaining these social relations reflected in the economic order.

The overarching question to think about in reading our world through these texts is the notion of social class. In the United States, for example, study after study has reported the likelihood of one's income to reflect that of her parents. This points to the notion of social reproduction and a lack of social mobility within the economic order. Other studies in the United States focus on the importance of occupational prestige as corresponding to social class. These studies indicate the executive elite (corporate CEOs, VPs, and board members) as having the most prestige, the professional class (doctors, lawyers, etc.) next, middle class (teachers, nurses, owners of small shops, middle management, skilled laborers), and finally working class (unskilled labor, service-sector employees, factory employees). I borrow these categories from Jean Anyon's (1980) article "Social Class and the Hidden Curriculum of Work," which investigates schools and social class. We will return to this article in detail in the next section, but for now it is helpful in furthering our understanding of social class as a social construct.

Anyon suggests three factors that describe one's place in the economic order of a capitalist society. The first, "ownership relations," points to the distinction among classes regarding the wealth that is owned. For example, Anyon points out that persons in the executive class own the majority of stocks in the United States. A recent Pew Research study indicates that the wealthiest 5% in the United States own 62% of stock. The poorest 60% own 4.2% of stock in the United States. The second characteristic determinant of social class, "relationships between people," suggests ones level of status in the occupational workforce. Think of the differing levels of authority between boss and employee and how these correspond to the class levels outlined in the previous paragraph. Those in the higher classes have more authority over others and autonomy in their work. The third factor determining social class is "relationship between people and their work." Think of the factory worker on an assembly line who has little understanding of the role his specific part plays in the grand scheme of the product being created as compared to the executive in charge of the factory operation who determines the entire process and knows its parts in total.

Social class: The social construction of class is the reproduction of a hierarchy of economic groups and general lack of social mobility. Social-class levels are marked by one's own wealth, level of authority versus obedience at work, and amount of autonomy at work.

The construct of social class is embedded within the capitalist economic order and it is important to understanding this context more fully. To do this, we begin with that most famous critique of capitalism by Karl Marx. In sketching the critique, I encourage you to take your own stance on economics, and we will think through some options at the conclusion of this section. Marx first published *Capital, Volume 1* in 1867. As a critique of capitalism, he works within the logics of capital to demonstrate its inherent inequality. In other words, Marx does not discuss the potential for corruption or excessive greed but describes how capitalism is structured to create inequality and necessitates the creation of class hierarchy.

First is Marx's point about the exploitation of labor in the market. He poses the following question: How is it that a capitalist can make a profit when he functions in a free market of equivalents? In the open market, it is presumed that goods and services are exchanged for materials at a price deemed fair to the two actors in the exchange. So how does profit turn up suddenly in the hands of the capitalist? Marx proposes that one value in the market serves two roles, and that is the labor value. On the one hand, a laborer receives money for her work. Let's call this amount W. This amount is negotiated based on how much the laborer thinks she needs to earn to support herself. Assuming for the moment this is the clothing industry, let's say our laborer earns \$10 per sweater she puts together. On the other hand, the laborer adds value to the products on which she is working; let's call this P. The capitalist buys the yarn needed for sweaters at \$20 per sweater but sells the sweaters for \$100. This means that the laborer has added \$80 of value, minus of course other costs that the capitalist has to pay such as the knitting needles, building where the knitters work, and so on. So P is, let's say, approximately \$50 per sweater. The laborer has added a value of \$50 to the materials but is only paid \$10. In this sketch, you see how P is greater than W, and this, Marx argues, is the exploitative situation by which capitalist profit emerges. As you can see, such an argument rests on mathematical knowledge to understand how capitalism exploits labor.

In order for the capitalist system to survive, there needs to be a consistent supply of laborers upon which the capitalist class can draw. Marx argues that the system reproduces itself, giving rise to class distinctions such as the capitalist class and working class.

> Capitalist production therefore reproduces in the course of its own process the separation between labour-power and the conditions of labour. It thereby reproduces and perpetuates the conditions under which the worker is exploited. It incessantly forces him to sell his labour-power in order to live, and enables the capitalist to purchase labour-power in order that he may enrich himself. It is no longer a mere accident that capitalist and worker confront each other in the market as buyer and seller. It is the alternating rhythm of the process itself which throws the worker back onto the market

> again and again as a seller of his labour-power and continually transforms his own product into a means by which another man can purchase him.
>
> *(Marx, 1990, p. 723)*

In the separation of the laborer from the means of production, the working class looks to the capitalist for means of survival. In most cases, the initiation of such relationships occurs alongside a surplus of people forced from their previous means of survival. For example, Marx describes the expropriation of peasants from their land in 15th-century England. As a result, ownerless individuals look to find paid employment to engage and survive within the market system they are thrust into.

Looking at more than a century of capitalist economic structures, neo-Marxist David Harvey describes the balance that can occur between the working class and the capitalist class. First off, the struggle and triumph of collective bargaining presents the working class's best efforts at reining in the exploits of the capitalist class. Here, workers form unions to negotiate wages collectively; doing so as a large group gives each laborer more strength in struggling against a capitalist whose interest is in exploiting labor. Many examples of unions show that collective bargaining effectively lowers capitalist rates of exploitation. Progressive nations have laws that support the creation and maintenance of unions. Harvey writes, "A 'class compromise' between capital and labor was generally advocated as the key guarantor of domestic peace and tranquility" (2005, p. 10). However, Harvey suggests how such compromise over exploitation only occurs when capitalists enjoy economic prosperity.

When capital power weakens, as in the United States in the 1960s, capitalists target the class compromise. Harvey and others describe this, the modern era, as neoliberalism, in which the political and economic elite aim to draw back compromises between capitalists and labor. The neoliberal era assumes the logics of a capitalist free market on all aspects of the ideal social life. In the capitalist free market, all individuals are assumed to behave rationally in their own self-interest. A capitalist makes the best decisions with respect to hiring labor, purchasing raw materials, and so on; a laborer sells his labor to the capitalist who will compensate the best, and so forth. In this world, everyone competes for greater profits and rewards. In neoliberalism, modern social life should assume such practices.

Public schools, for example, should compete for students; such competition would engender in schools their pursuance of self-interest and, consequently, wise decision making. As Harvey notes, the neoliberal era requires a state to simultaneously support the free market and get out of its way:

> Neoliberalism is in the first instance a theory of political economic practices that proposes that human well-being can best be advanced by liberating individual entrepreneurial freedoms and skills within an institutional

> framework characterized by strong private property rights, free markets and free trade. The role of the state is to create and preserve an institutional framework appropriate to such practices. The state has to guarantee, for example, the quality and integrity of money. It must also set up those military, defense, police, and legal structures and functions required to secure private property rights and to guarantee, by force if need be, the proper functioning of markets. Furthermore, if markets do not exist (in areas such as land, water, education, health care, social security, or environmental pollution) then they must be created, by state action if necessary. But beyond these tasks the state should not venture. State interventions in markets (once created) must be kept to a bare minimum because, according to the theory, the state cannot possibly possess enough information to second-guess market signals (prices) and because powerful interest groups will inevitably distort and bias state interventions (particularly in democracies) for their own benefit.
>
> *(Harvey, 2005, p. 2)*

In the neoliberal era, public schools should function for the needs of the free-market system.

You may be wondering if teaching with a critical perspective requires you to be anticapitalist. This is certainly one option you can take, and I encourage you to learn about alternative economic structures that might take its place, such as anarchist theories of economics. The concept of anarcho-syndicalism is a good place to start, and you can read Noam Chomsky's (2005) introductions on the topic (see reference list at the conclusion of this chapter). Alternatively, learning of the exploits inherent in capitalism can position you better within the *realpolitik*, in which you examine the status quo critically and push for what changes can be made given the current conditions and circumstances and toward a more fair class compromise. For this option, a sort of reining in of capitalism, we can turn to John Maynard Keynes and what is now known as Keynesian economics. This is similar to the balance between working and capitalist classes that we learned of earlier through David Harvey's work. In Keynes's 1936 *The General Theory of Employment, Interest and Money*, he argues that the free market cannot maximize employment for the working class (Keynes, 2009). Left to its own devices, there will exist significant unemployment and poverty. This implies the need for government interventions and job creation. Contemporary economists promoting this view include Robert Reich and Joseph Stiglitz. So the simple answer is that teaching with a critical perspective requires you to understand the exploitation inherent in capitalist logics and think through the range of possibilities, from alternative economic structures like anarchism and communism to more progressive capitalist nations. In the next section, we think about how the capitalist construction of social-class hierarchies relates to education.

Class Reproduction, Schools, and Critical Pedagogy

Reading about the exploits of capitalism and the lack of social mobility within the system may cause you to wonder why individuals and groups, especially those in lower social classes, do not fight to change it more often than they do. The answer to this question lies with a look at government-provided education. Contrary to what neoliberal thinking will tell you, capitalism requires government structures for a few key components. The enforcement of a law on private property is a clear example. Without a government-sponsored and enforced law of ownership, capitalists cannot own the means of production and thereby exploit laborers. Additionally, government-run public schools reproduce social-class hierarchies necessary for the reproduction of social-economic hierarchies. In this section I will briefly review both theories on and observations from public schools, especially with reviews of contributions by Pierre Bourdieu and Jean Anyon. In addition, we will review a significant strand of thought regarding teaching against the reproduction of class hierarchy: the Marxist-inspired critical pedagogy.

Schools prevent mass uprising from the populace because they engender a feeling that we live in a meritocracy. Such an ideal describes a society in which individuals in positions of power, authority, and wealth have achieved their status as a result of merits they have attained. In other words, working hard in school leads to success in life. Some of us proudly display our success as a school student in order to demonstrate our qualifications for a role in society, be it a managerial position, political office, what have you. However, theorists who recognize the necessity of social class for the economic order question meritocracy, claiming that its mythology leads to self-blame for one's lack of achievement. Those of us not doing well in school and subsequently attaining roles in society deprived of power and wealth blame ourselves for inadequate brains and/or poor work ethic. To consider public education as a servant for capitalist society requires that we consider meritocracy as a myth: How does public education perpetuate and reproduce social class rather than allow for social mobility? It turns out there are two related answers. First, school culture and its curriculum more closely match the culture and experience of upper- and upper-middle-class life, thus putting students from higher-social-class backgrounds at an advantage. Second, schooling experiences are differentiated according to social class, with students in lower social classes receiving a schooling program that relates to lower-social-class work expectations, and, similarly, students from higher-social-class backgrounds receiving an education in what is required to work in higher-class careers.

For the first, we look to the French philosopher and social theorist Pierre Bourdieu. In a chapter titled "The Forms of Capital" (1986), he describes three concepts of capital: economic, cultural, and social. Economic are those assets that can be immediately converted into material wealth, cultural as aspects to an individual that can be converted into economic capital, for example in the form of knowledge from education, and social, or the networks of "who you know" that can be converted into economic capital. Of the three, cultural capital relates most to the educational process. Think of cultural capital as the knowledge, behavior, and habits of mind that reflect one's status in the economic order. Essentially, social reproduction works through the schools via the notion of cultural capital because schools reflect the culture of the upper and upper middle classes. Thus, a student from the upper and upper middle classes will learn this culture both at school and elsewhere in life—at home, for example. The student has advantage in the school setting, whereas a working class student does not. When it comes to measures of school performance, it is inevitable that the upper- and upper-middle-class students on the whole outperform the lower-class students. As Bourdieu suggests, the stratification of "succeed and not succeed" reproduces the social classes and limits social mobility.

Cultural capital: As the knowledge, behavior, and habits of mind for success in the economic order, this is what is taught and assessed by schools. Upper-class students have an advantage because they learn it both in and out of school.

For the second way that schools reproduce social class, we look to educational scholar Jean Anyon. Instead of merely providing everyone a similar education in which upper class students have advantages, Anyon documents that schools are differentiated by economic class and provide differing instruction. Her 1980 article "Social Class and the Hidden Curriculum of Work" describes her research in New Jersey schools that are stratified according to the following social classes: executive elite, affluent professional, middle class, and working class. All schools used the same textbooks and similar curriculum. However, her observations of the instruction demonstrated stark contrasts. In the working-class schools, work is procedural and textbook oriented, usually involves practice worksheets, and generally follows directives from an authoritarian teacher. Contrast that with the executive elite school, where work is characterized as developing analytical thinking, debating opinions on social issues, writing original essays, and having some freedom to pursue individual interests. Anyon also noted distinctions regarding the attitude with which teachers approach their students. She concludes that a "hidden curriculum" is at play, in which students learn their relationships to others as dictated by their place in the social economic order. For example, the obedience to authority required by factory workers is what is taught in working-class schools.

Anyon and others have provided both theoretical and empirical arguments about the social reproduction of schools. Classic books on the topic include *Schooling in Capitalist America* by Herbert Bowles and Samuel Gintis (1977) and

Learning to Labor by Paul Willis (social reproduction in the UK). Essentially, schools facilitate success of upper-class students in two ways: by matching the curriculum to upper-class culture and by offering different schooling experiences to students in the differing social classes.

Having understood this critique of schools in the ways they reproduce social-class hierarchy, as critical teachers of mathematics, we need to look at theories of practice that will interrupt such social reproduction. Critical pedagogy, which now comprises a host of subfields, emerged initially as a Marxist interpretation of the role teaching and learning can play in raising class consciousness, a central feature of Marx's call to action in disrupting social-class hierarchy. Initial work in developing the concept is typically credited to Brazilian educator Paulo Freire.

Freire's *Pedagogy of the Oppressed* (2000, first published in 1968) has taken on great meaning for teachers with a critical perspective. Essentially, Freire describes a naïve consciousness in which typical ideas are held about groups of people and the way society functions, and especially those ideas that keep power structures as they are. Working in Brazil and elsewhere in Latin America, Freire's practical work involved the oppressed working class who internalized a negative image given the social-class relations set forth by the upper class. Freire led the practice of teaching literacy through codifications, or images depicting an individual's position in the political-economic order. Images prompted discussion and eventual participation in the written word, all motivated by the learner's development of a critical consciousness that replaces the false naïve consciousness. In Portuguese, Freire gives the term *conscientizacao* to describe such an awareness of new consciousness.

Critical pedagogy: A theory-informed practice (praxis) in which teaching and learning emphasizes raising consciousness about social-class relations. The concept has expanded and been applied to pedagogies of raising consciousness about many if not all forms of oppression beyond social-class hierarchy.

The idea of raising consciousness has come to be relevant for each and every one of the chapters of this book. The concept of teaching to raise consciousness, as described earlier, can be applied to race and gender constructs just as it can be useful for teaching class consciousness. Given its consistent thinking with Marxist theory (false consciousness) and Freire's initial goals for Latin America, I introduce it here in the chapter on class and education. His pedagogy increased awareness of the oppressive relationship in the economic order and motivated revolutionary thinking about economics and social relations. Many have followed this tradition, including several scholars in the United States who were students of Paulo Freire. These include bell hooks, Peter MacLaren, Donaldo Macedo, and Antonia Darder.

This section highlighted major discussions on class and education. On the one hand, we have a great many studies and concepts that criticize education for its role in social reproduction of economic class. On the other are important contributions from Freire and others that teach us how to interrupt production. Seeing these contributions in two camps, then, as mathematics educators, we must ask ourselves the following questions: (1) What elements in my practice of mathematics education reinforce class structure and limit social mobility? (2) What efforts can I make to change these elements, or perhaps more impactful, more explicitly teach to raise consciousness when teaching mathematics? The next section reviews important considerations on these two questions.

Social Class and Mathematics Education

With broader understandings now at hand, this section digs deeper into the thinking on social-class relations and mathematics education. Similar to the previous section, there are contributions regarding how mathematics education reproduces social-class relations as well as efforts to directly interrupt reproduction. As for the reproduction of social classes, we need to think deeply about what school mathematics is. How does school mathematics reflect upper- and middle-class culture and thus give advantage to students from higher social classes? How are mathematics education programs differentiated according to social classes, where some students are offered a different version of teaching and learning than others? Finally, a host of direct applications of Freirian theory and practice to mathematics education will be discussed.

Researchers across the globe have produced empirical evidence for mathematics education's role in reproducing social-class relations. The article "Structural Exclusion Through School Mathematics: Using Bourdieu to Understand Mathematics as a Social Practice" by Robyn Jorgensen, Peter Gates, and Vanessa Roper provides case studies documenting how mathematics education effectively includes and excludes students based on their social-class backgrounds. As reflected in the title, they draw on Bourdieu's notion of cultural capital and, more specifically, the related concepts of habitus and field. The *habitus* is the taken for granted explicit and implicit rules, assumptions, and habits of mind of cultural capital. In general, the habitus of students from higher-class home lives matches the habitus of school life. Pertaining to mathematics, the habitus regards it

> as important—as a gatekeeper to travelling successfully through the educational system and as an inherent marker of intellect . . . Mathematics acts as a marker of success in schools, and consequently, mathematics is a useful context in which to explore the inequality apparent in the education system as a whole because it performs a role of social segregation.
>
> *(Jorgensen et al., 2013, p. 223)*

As for the *field*, the authors suggest mathematics education as a key organizer of social reproduction, in which

> various individuals interact, and these encounters produce the accepted social practices that typify the field . . . For example, the mathematics curriculum is structured in a particular way that privileges certain forms of thinking, pedagogy is structured to distinguish between different learners, expectations become organized around visions of different futures, behaviors are shaped around the image of the ideal pupil, relationships with parents place teachers in very specific positions of authority and so on . . . Mathematics holds a privileged place in the school curriculum, and ability in mathematics is highly prized and valued, defining learners as "can do" or "can't do."
>
> *(p. 224)*

The authors of the article provide examples showing that students from the UK who come from higher-class homes have advantages in learning school mathematics. For example, in the early years, parents and caregivers "play school" and ask "school-like" questions more readily than their lower-class counterparts. This does not imply that lower-class parents and caregivers are less able or smart to do so and instead highlights the cultural mismatch between their homes and school life and congruence between higher-class homes and school life. As students work through the schooling system, opportunities to engage in learning the habitus differentiate through structures like tracking and are further reinforced by differences in advantage back at home. As shown by the case studies by Jorgensen and colleagues, students from higher-class homes might have experiences with the habitus, the behaviors and habits of mind required by mathematics education, whereas lower-class students might not.

Providing research in the United States context, Sarah Theule Lubienski (2000) provides an interesting discussion regarding social class and reform mathematics. Recall the push for mathematics teaching and learning that emphasizes student-centered learning, discovery, problem solving, manipulatives, and so forth, as discussed in Chapter 2, and the structure for lesson planning that requires students to experience an activity and then debrief in whole-class discussion. Lubienski's work requires we reconsider these strategies carefully for how they might reinforce social-class reproduction. Her article "A Clash of Social Class Cultures? Students' Experiences in a Discussion-Intensive Mathematics Classroom" draws on interviews and observations from a reform mathematics classroom to conclude that some aspects, specifically the teacher prompts that lead to whole-class discussion, more readily match the home life of students from higher-class backgrounds. Again, it's a match of the habitus between home and school mathematics. Lubienski's work describes how the open-ended, abstract-thinking questions required by reform efforts were less familiar for lower-class students, who instead preferred questions with answers to specific, contextualized problems. This does not suggest that lower-class students cannot do the other types of higher-order-thinking questions but that the teacher must be aware of how she structures these opportunities and consistently reflect on whether students are being advantaged/disadvantaged in her classroom due to social class.

These empirical studies give support to the specifics of a school mathematics that reproduces social-class hierarchy. More theoretical arguments relate the typical mathematics education program to economic relations in capitalism. Education policy scholar Michael Apple (1992) highlights the role that mathematics plays for its "administrative and technical relevance" to capitalism:

> The accumulation and control of technical/administrative knowledge by a limited group of people is essential in our science-based industries and for the production of material and weapons systems for the Department of Defense. In the calculus of values we use to sort out "important knowledge" from "less important knowledge," business and industry, as well as the government, place a high value on knowledge that is convertible ultimately into profits and control. The possession of such knowledge through patents, hiring practices, funding research institutes, and so on is crucial in the highly competitive economic system we live in. But by possession here I do *not* mean that you and I should have such knowledge or that it be widely distributed. Rather, I mean that those with economic, political, and cultural power can employ it in ways they see fit.
>
> *(p. 420–421)*

Apple reminds us that mathematical knowledge is essential in defense and science-based industries, and the control of this knowledge must be strict in order to maintain economic relations of power.

Yet another contribution in this regard comes from the Danish mathematics education theorist Ole Skovsmose. Rather than prioritizing the content of school mathematics itself, as Apple does, he considers the experience of school mathematics and its relevance to social-class relations. In the following excerpt from his book *An Invitation to Critical Mathematics Education* (2011), Skovsmose describes how school mathematics encourages a "prescription readiness" in which students are taught to obey authority:

> From the perspective of understanding mathematics many regulations and corrections, so characteristic of the school mathematics tradition, appear irrational. However, when students have been directed through the 10,000 exercises, they might have learnt something which need not have much to do with any mathematical understanding. Their learning might crystallize into a prescription readiness . . . These exercises seem to take the form of

> a long sequence of instructions. Could it be that school mathematics tradition cultivates a prescription readiness, which prepares the students for participating in work processes where a careful following of step by step instructions without any question is essential? Could it be that such a prescription readiness is serviceable for very many job functions in our society and that the school mathematics tradition serves society perfectly well in exercising this readiness? Could it be that a prescription-readiness, including submission to a regime of truths, cultivates socio-political naivety and blindness that is appreciated at today's labor market? Could it be that a prescription-readiness fits perfectly well the priorities of a neo-liberal market, where hectic and unquestioned production serves economic demands?
>
> *(pp. 9–10)*

Thus we have several critiques highlighting mathematics education's relevance to social-class hierarchy and the reproduction of social classes. Moving from critique to opportunity, there exist a host of contributions in mathematics education that directly contradict the social reproduction of classes. Many of these draw on the writings and practice of Paulo Freire, discussed in the earlier section. In the early 1980s, Marilyn Frankenstein applied critical pedagogy to mathematics education both theoretically and in practice, especially in her work teaching basic-skills mathematics courses for adult learners. Her 1983 article "Critical Mathematics Education: An Application of Paulo Freire's Epistemology," is pointed to as a clear introduction to teaching mathematics that interrupts typical social-class relations. She focuses on the problem-posing nature of Freire's approach to raising consciousness and suggests mathematics' role in social-class relations and mathematical knowledge as key to understanding it. For example,

> A mathematically illiterate populace can be convinced that social welfare programs are responsible for their declining standard of living, because they will not research the numbers to uncover that "welfare" to the rich dwarfs any meager subsidies given to the poor. For example, in 1975 the maximum payment to an Aid for Dependent Children family of four was $5,000 and the average tax loophole for each of the richest 160,000 taxpayers was $45,000.
>
> *(p. 327)*

Frankenstein's work catapulted efforts to teach mathematics for liberation in the Freirian tradition. Although many others can be included here, a comprehensive study in these efforts is Eric Gutstein's (2006) *Reading and Writing the World With Mathematics*. Here he provides several lesson examples of teaching mathematics to interrupt social-class hierarchies. These examples are richly detailed with documentation of urban, low-income student experiences with such a curriculum. The book's countless examples inspire us for the successes we can have in teaching mathematics that raise consciousness. As I stated in the preface to this book, most if not all of the constructs (race, class, gender, etc.) intersect significantly. In a way, separating them by chapter does a disservice to the importance of intersectionality, and we attend to this point in the final chapter where we "put it all together." Gutstein's contribution details the intersections specifically of race and class and the role that typical mathematics education plays in this as well as the ways that an alternative mathematics education can push against it. I chose to introduce Gutstein's contribution here given Freire's Marxist origins, but I could have just as easily discussed the work in the previous chapter on race. The book provides several examples of the varieties of considerations and approaches to teaching mathematics that interrupt social-class hierarchies as well as racial injustice.

In concluding this section on mathematics education's relationship to social class, we can revisit the major considerations put forth by the research. First, we have to consider the ways that students from higher-class backgrounds are given advantage in mathematics education. This could be because home life corresponds more readily to the mathematics education classroom in a number of ways as well as the differentiated teaching and learning experiences in mathematics that are afforded to students from differing social-class backgrounds. Second, schools privilege mathematics instruction because its content and teaching/learning experiences correspond to social-class relations. Mathematical content is necessary for science-based industries, and the typical teaching/learning of mathematics reinforces obedience to authority. Finally, we have much to be inspired by with the applications of a Marxist pedagogy to mathematics instruction. In interrupting social-class reproduction, we are called to a careful study of the teaching and learning done by Frankenstein and Gutstein.

Activities and Prompts for Your Consideration

1 Use the frameworks for teaching and learning mathematics that were introduced in the first chapter and your institution's lesson plan format to create a mathematics lesson in the Freirian tradition. This lesson must teach both a content and process standard suitable to your grade level and provide an experience through which students will begin to think critically about social-class relations. An example scenario is suggested if you are having trouble getting started:

Consider writing a lesson that teaches students the Marxist concept of exploitation of labor (discussed in the first section of this chapter). To explore this notion, students will be required to employ both number and operations thinking and algebraic thinking. The way you set up the example problems and scenarios will depend on the grade level and student readiness. On the lower end, use more examples with numbers and have students compare and contrast whether one capitalist exploits more than another. On the higher end, employ algebraic thinking earlier in the lesson and require students to debate algebraically defined rules that curb a capitalist's rate of exploitation. As an example, students might come up with: the amount paid to the laborer cannot be less than 2.5 times the added value the laborer provides to the product. Give students multiple fictional examples of a capitalist's labor costs, material costs, and means of production upkeep. Ultimately, all students will have discussions about what is fair, and some might question whether capitalists should own the means of production at all or if perhaps all laborers should collectively own and share profits equally.

2 Discussions with peers:

(a) Reflect on your own schooling experience. Would you say that your grade school was homogenous or diverse with respect to social class? What kind of teaching and learning, generally and specific to math, took place there? Compare and contrast with your peers to determine the extent to which you think schools and school mathematics reproduce the social-class hierarchy.

(b) After reading this chapter and given your own experiences, where do you stand with respect to capitalism? For example, do you think there should be a balance between the social classes, and is this the responsibility of the government? Or do you think the government and/or economic system should be replaced with something different, like anarchism or communism?

3 Work with a partner to interview two students about their perceptions of "math class." Ask questions related to their perceptions of what mathematics is and what is expected of them during instruction. You want to get the sense of whether students feel that they must obey rules or whether they are encouraged to think critically, communicate, and justify their mathematical reasoning. With your partner, compare and contrast the two interviews, noting how they might correspond to the notions of social class related to this chapter. Be careful not to make assumptions about your interviewees' social-class backgrounds (and I do not suggest that you specifically ask them about this). As you discuss and reflect, consider ways that teachers can interrupt the ways that mathematics education reproduces social-class hierarchies.

References

Anyon, J. (1980). Social class and the hidden curriculum of work. *Journal of Education 162* (1): 67–92.

Apple, M. (1992). Do the standards go far enough? Power, policy, and practice in mathematics education. *Journal for Research in Mathematics Education 23* (5): 412–431.

Bourdieu, P. (1986). *The forms of capital.* Available at https://www.marxists.org/reference/subject/philosophy/works/fr/bourdieu-forms-capital.htm.

Bowles, S. & Gintis, H. (1977). *Schooling in capitalist America: Educational reform and the contradictions of economic life.* New York: Basic Books.

Chomsky, N. (2005). *Chomsky on anarchism.* Oakland, CA: AK Press.

Frankenstein, M. (1983). Critical mathematics education: An application of Paulo Freire's epistemology. *Journal of Education 165* (4): 315–339.

Freire, P. (2000). *Pedagogy of the oppressed.* New York: Bloomsbury Academic.

Gutstein, E. (2006). *Reading and writing the world with mathematics: Toward a pedagogy of social justice.* New York: Routledge.

Harvey, D. (2005). *A brief history of neoliberalism.* New York: Oxford University Press.

Jorgensen, R., Gates, P. & Roper, V. (2013). Structural exclusion through school mathematics: Using Bourdieu to understand mathematics as a social practice. *Educational Studies in Mathematics 87*: 221–239.

Keynes, J. M. (2009). *The general theory of employment, interest and money.* New York: Classic Books America.

Lubienski, S. (2000). A clash of social class cultures? Students' experiences in a discussion-intensive seventh-grade mathematics classroom. *The Elementary School Journal 100* (4): 377–403.

Marx, K. (1990). *Capital Volume 1.* New York: Penguin.

Skovsmose, O. (2011). *An invitation to critical mathematics education.* Rotterdam, Netherlands: Sense.

5

RATIONALISM, MASCULINITY, AND THE "GIRL PROBLEM" IN MATHEMATICS EDUCATION

Are there biological differences between men and women that require they be educated differently? Why do girls outperform boys in mathematics in the earlier grades, but as adults, boys are more likely to end up in mathematically intensive careers? How do mathematics teachers perceive students who are successful and not successful in these classes, and why do they give differing responses when discussing boys versus girls? These and other questions will be explored in this chapter, where we first trouble gender by affirming that these categories are social constructions rather than biologically based realities. Next, I apply these examined perspectives on gender to education in general with a look at important contributions from the literature. Finally, I review research in mathematics education that takes up these themes and calls us to examine our teaching practice from these perspectives.

Here is a brief, somewhat recent news event to motivate our work in this chapter. Lawrence Summers is a world-renowned economist who in the late 1990s served as U.S. President Bill Clinton's Secretary of the Treasury. Over the years, he has gained accolades for his work as an academic and in his service to banks and hedge fund management companies. In 2005, when he was the president of Harvard University, he made a statement at an economics conference that shook the mathematics and sciences world. Summers gave a speech highlighting the gender imbalance that greatly favors men over women in mathematics and science departments at colleges and universities. He pointed to a variety of possible explanations but suggested that he favors differences in cognitive ability over discrimination and cultural factors. To prove the point, Summers referenced difference in aptitude tests administered in the 12th grade. Needless to say, his speech caused quite a stir and is possibly the reason that the following year Summers stepped down as Harvard's president. This story begs you to think through the validity of Summers's claim. For example, to what extent can such test results accurately document a distinction in aptitude according to gender? To begin, we need to take a step back and think through what we mean by gender. We will start here in the first section of this chapter.

Gender and the History of Feminism

To start, take the distinction between biological sex and gender. Biological sex refers to one's genetic makeup, physical anatomy, and hormones and includes the categories male, female, and intersex. On the other hand, gender describes the typical behaviors, ways of being, and other characteristics that are thought to be masculine and feminine. A transgender person feels and exhibits the characteristics of the gender that is opposite their biological sex. The term "cisgender" describes people who express and feel the characteristics of the gender that aligns with their biological sex. Transsexual people are those that undergo hormone therapies and/or surgeries to reassign their biology to align with the gender they feel and are expressing.

Reviewing these examples helps clarify that gender is something we perform rather than a natural given fact. Advanced thinking on gender, in particular that coming from a critical perspective, suggests that we are nurtured to perform our gender rather than born to behave a certain way. Not just about blue for boys and pink for girls, the characteristics associated with masculinity and femininity are taught to us early on and continuously throughout life by the family, media, schools, our workplaces, religion, and so on. For example, masculine traits include competitiveness, strength, rationalism, anger, courage, assertiveness; feminine traits include dependence, emotionality, weakness, quiet, grace, and nurturing. As we grow up we are encouraged, some would say forced, to perform the traits aligning with the biological sex we are assigned. For many whose biological sex does not align with their gender, this is an extremely painful process. And taken in total, the societal expectation of such gender roles has significant consequences, including the oppression and lack of opportunity for girls, women, transgender individuals, and transsexuals. We will be looking specifically at how this operates in the mathematics education space.

The previous paragraphs introduce the concept of gender as social construct, and I encourage you to explore these further by reading classic texts on the subject. Judith Butler is a gender and sexuality theorist with many books on the topic. Her book *Undoing Gender* (2004) provides deeper theoretical considerations on several questions related to gender. For example, consider the following discussion regarding the intersex, or people who are born with ambiguous physical characteristics with respect to sex:

> The question of surgical "correction" for intersexed children is one case in point. There the argument is made that children born with irregular primary sexual characteristics are to be "corrected" in order to fit in, feel more

> comfortable, achieve normality. Corrective surgery is sometimes performed with parental support and in the name of normalization, and the physical and psychic costs of the surgery have proven to be enormous for those persons who have been submitted, as it were, to the knife of the norm. The bodies produced through such a regulatory enforcement of gender are bodies in pain, bearing the marks of violence and suffering. Here the ideality of gendered morphology is quite literally incised in the flesh.
>
> *(p. 53)*

With a discussion that "troubles gender" at hand, we now can think about how such constructions of femininity and masculinity relate to power structures throughout history. By ascribing gender roles and their characteristics to persons with particular physiologies, groups of people, specifically women and the transgendered, endure oppressive institutional structures and societal habits of mind positioning them as "less than" or inferior. One way to think through power relations and gender is by reviewing the history of the feminist movement. The movement is typically described in waves. First-wave feminism refers to the suffrage movements in several countries during the later 19th and early 20th centuries. Second-wave feminism refers to the rebirth of gender-equality discussions of the 1960s to 1980s, including a focus on sexuality, family life, inequality at work, and unequal educational opportunities. In the 1990s, a "third wave" or "postfeminism" more strongly integrates queer and nonwhite perspectives in feminism as well as strong opposition to gender norms and binaries.

A concise and accessible review of the history of feminism comes from bell hooks, a Black feminist, in her text *Feminism Is for Everybody*. She opens with some clear responses to the common misunderstanding of feminism, which she characterizes as follows:

> I tend to hear all about the evil of feminism and the bad feminists: how "they" hate men; how "they" want to go against nature and god; how "they" are all lesbians; how "they" are taking all the jobs and making the world hard for white men, who do not stand a chance. When I ask these same folks about the feminist books or magazines they read, when I ask them about the feminist talks they have heard, about the feminist activists they know, they respond by letting me know that everything they know about feminism has come into their lives thirdhand, that they really have not come close enough to the feminist movement to know what really happens, what it's really about. Mostly they think feminism is a bunch of angry women who want to be like men. They do not even think about feminism as being about rights—about women gaining equal rights. When I talk about the feminism I know—up close and personal—they willingly listen, although when our conversations end, they are quick to tell me I am different, not like the "real" feminists who hate men, who are angry. I assure them I am as

> real and as radical a feminist as one can be, and if they dare to come closer to feminism they will see it is not how they have imagined it.
>
> *(2014, p. vi–vii)*

Whereas feminism is so often misunderstood, hooks reminds us that the term simply rejects that men are superior to women. This is not altogether different from an antiracist stance rejecting that whites are superior to nonwhites.

One feature of hooks's review on the history of feminism is her attention to the emergence of feminism within and across other avenues of social justice. For example, in the civil rights movement and in movements for social class equity, hooks notes that women highlighted the expectations that they follow male leaders in these social movements. The targeted messages and actions of such movements for equality provoked inconsistent thoughts and feelings for the women who took up such causes. The problem was with "men who were telling the world about the importance of freedom while subordinating the women in their ranks" (p. 2). As hooks describes the history, early on, feminists polarized amongst those that sought more sweeping social reforms aimed at identifying consistencies among oppression (racism, classism, sexism) and eradicating these together and those that targeted the women's cause specifically with goals for equality in schools and the workplace. The latter had greater success, especially with the

> **Feminism:** A movement spanning more than 100 years that aimed first to establish equality for women through suffrage (first wave), equality in home and workplace (second wave), and finally the undoing of gender norms as well as highlighting intersections to race, class, and other social identities (third wave).

corporate powers that be and emerging conservative responses to the civil rights era already in place. These less radical feminist efforts, as hooks argues, contributed to the stalled progress of the civil rights movement:

> We can never forget that white women began to assert their need for freedom after civil rights, just at the point when racial discrimination was ending and black people, especially black males, might have attained equality in the workforce with white men.
>
> *(hooks, 2014, p. 4)*

She notes the embrace of the less radical version of feminism by most women and the location of more radical feminist thinking as restricted to academic circles. Referring to the more mainstream version as "lifestyle feminism," hooks notes that such conversation and action lacks its political origins. The central aim of

Feminism Is for Everybody is to regain this, especially by highlighting the interplay between sexism and related oppression, including race and social class. As we shall see, a host of approaches exist in the work on gender and mathematics education and reflect a similar range of stances.

Recall the opening discussion regarding the social construction of gender. These conversations emerge with more recent theoretical work in feminism that seeks to displace the feminine subject at its center. Third-wave feminism, as a poststructural feminism, calls into question the concept of gender, as Judith Butler clearly does in *Undoing gender*. As well, third-wave feminism provides direct responses to hooks's concerns with the first and second waves. Third-wave feminism highlights the intersection of sexism with, for example, racism, given the writings of Black feminists, such as hooks herself. Drawing on poststructural theory, third-wave feminism is noted for its work with Foucault and his contemporaries, as well as queer theory. We will take a look at both in the next section, a brief discussion of sexuality.

The *History of Sexuality*

Like gender, sexuality is based more on social norms and expectations than on biology. Although this social identity deserves its own chapter in the book, unfortunately there is not enough scholarship in mathematics education on this topic. However, I find it important and necessary to think through as we consider teaching mathematics with a critical perspective, and I placed it within the gender chapter because a critical understanding of sexuality motivates a similar critical conception of gender, and there are excellent applications of this to mathematics education. In this brief section, we review the considerations of famed social theorist Michel Foucault.

Foucault's work contributes significantly to the poststructuralist perspective on social life. He examined a variety of topics, from mental health to knowledge to imprisonment, by consistently describing how what is at one historical moment perceived to be normal and "a fact" is actually a constructed reality that represents a range of historical confluences, from politics to economics and other social relations. There exist "discourses" and "regimes of truth," what we might take as given, for granted, always-having-been-true ideas that actually shape our relations to one another and significantly impact one's place within a web or flow of societal power. We can conceive of sexuality, gender, and even mathematics as such regimes of truth. Working among a variety of topics, Foucault describes discursive formations, or the processes that herald such discourses as regimes of truth. His work and that of his peers framed much of social theory taking place in the 1990s and beyond, all of which loosely corresponds to a poststructuralist theory. Although this is difficult to define and many so-called poststructural theorists actually dislike the term, what seems consistent among these thinkers are their rejection of unifying theories that explain all social life (such as Marxism) and their efforts to consider multiple perspectives coming to bear on individuals and institutional circumstances.

One of the topics addressed by Foucault is the notion of sexuality in his book first published in English in 1970: *The History of Sexuality: An Introduction Volume 1* (1990). The book describes the discursive formation of sexuality at the end of the 19th century and continuing to the present day. Foucault discusses the relations between an uprising bourgeoisie and capitalist economics and the beginnings of repressed sexuality. Examples include newfound legal and practical considerations claiming the reproductive role of sexuality as paramount. These newer circumstances invented new identities fitting outside this definition, such as the homosexual, tainted by claims of perversion. In spite of such negative distinctions, we make progress throughout history as, for example, with movement toward lesbian, gay, bisexual, transgender, queer (LGBTQ) rights. However, these triumphs exist within such categories and definitions and sometimes may reinforce the superior/inferior binary at the root of the problem. Foucault writes:

> There is no question that the appearance in nineteenth-century psychiatry, jurisprudence, and literature of a whole series of discourses on the species and subspecies of homosexuality, inversion, pederasty, and "psychic hermaphrodism" made possible a strong advance of social controls into this area of "perversity"; but is also made possible the formation of a "reverse" discourse: homosexuality began to speak in its own behalf, to demand that its legitimacy or "naturality" be acknowledged, often in the same vocabulary, using the same categories by which it was medically disqualified.
>
> *(p. 101)*

Take, for example, contemporary progress of LGBTQ movements that have successfully pushed against the original claims that homosexuality is perverse or disorderly. Some of the fuel for these successes has been to articulate queerness as biologically based. However, Foucault claims, as mentioned earlier, that such success came as the result of claims that further define homosexuality as an "other." Typically, poststructuralism argues that advancement comes with more fluid understandings of binaries rather than strict definitions that relate in some way to a "provable" scientific fact such as genetics. In this way, the claim that a homosexual was "born this way" continues to reinforce an objectifying perspective. Similarly, Foucault would argue that marriage equality acclimates the "other" into a sanitized, repressive sexuality. This is not to say that these conversations and successes are worthless in fighting the regime of truth that has cast nonheterosexual activity as inferior and disorderly.

This brief discussion of sexuality and Foucault's theories serves to highlight the need for more critical work in mathematics education but also to advance our understanding of social constructs as these contributions have been applied to other social identities, including the primary goals of this chapter in troubling gender. As we shall see in the next section, discussions of gender in education began with discussions that move us forward but continue to reinforce gender binaries as they have been constructed. In applying Foucault to gender, we will see that queer theory moves us forward in deconstructing binaries. In at least one case, queer theory has been applied to theorize mathematics education and the notion of good mathematics students being "born this way." We will take a look at this as one example in the final section, but first, we need to see the more broad applications of feminism made toward pedagogy and general education.

Gender and Education

With a discussion of gender and sexuality as social constructs underway, we turn this onto education and schools in general with a review of a handful of important sources in the field. This will properly situate our review of what work has been done with respect to mathematics education and gender and how teaching mathematics with a critical perspective can more properly reflect advanced understandings of gender and sexuality. The first of these do not fully position us to teach as critically as we can but nevertheless provide an initial departure point for moving us in this direction.

One seminal text that certainly got several conversations going is *Women's Ways of Knowing: The Development of Self, Voice, and Mind* by Mary Field Belenky, Blyth McVicker Clinchy, Nancy Rule Goldberger, and Jill Mattuck Tarule (1986). The authors summarize their research on 135 women from varying social class, educational, and race/ethnicity backgrounds to describe five developmental positions characterizing the ways that a woman can understand, interpret, apply, and critique knowledge. The categories run from a "silenced knowledge" to a "constructed knowledge" and characterize the knower's relationship to the knowledge. This includes how she was taught knowledge, what she finds as more and valuable forms of knowledge, and whether she sees ideas as static or dependent on context. For example, the third developmental position is titled "subjective knowledge," in which the learner has distrust for analytical ways of knowing in favor of more concrete thought processes rooted in experience. To contrast, the highest position, "constructed knowledge," considers all forms of knowledge, including both analytic and experience-driven ideas, primarily with the clear acknowledgment that an individual's knowledge is constructed by her experiences and analysis. The constructivist knower listens to all perspectives and would begin a respond to a question with, "My understanding is . . ."

Primarily, Belenky and colleagues provide a dialogue in response to theories of knowledge that up until then were dominated by male-centric frameworks. As we will see in the next section, *Women's Ways of Knowing* has been directly applied to mathematics education as well as several other disciplines. As well, the work has been perceived more deeply by third-wave feminists, who take up the post-structural standpoints introduced in the previous section on sexuality. Although

Women's Ways responded directly and necessarily to a male-dominated viewpoint on psychology, it reinforces the gender binaries that the third wave aims to destabilize by perceiving women in a unitary manner. Thus we need to continue our study by looking to the work of poststructural feminist applications to education.

Carmen Luke and Jennifer Gore situate their collection of scholarly writings on feminist pedagogy (*Feminisms and Critical Pedagogy*, 1996) within poststructural feminism: "That is, texts, classrooms, and identities are read as discursive inscriptions on material bodies/subjectivities. Pedagogical encounters and pedagogical texts are read both as a politics of signification and as historically contingent cultural practice" (p. 4). The collection of scholars in this work particularly thinks through critical pedagogy (discussed in detail in Chapter 4) as a "regime of truth" itself containing particular discourses that fail to fully challenge gender binaries and hierarchies.

> Hence, our poststructuralist feminist task is to go beyond the deconstruction of the normative masculine subject valorized as the benchmark against which all others are measured, and to examine how and where the feminine is positioned in contemporary emancipatory discourses (including feminist discourses). The high visibility of "gender" in social justice and equity programs and policies, and its status in almost all progressive pedagogical tracts, easily obscures the fact that equal space and representation in curriculum, policy or the conference agenda does not in itself necessarily alter the status of the feminine as an add-on category or compensatory gesture. As such, the poststructuralist feminist agenda remains focused on challenging incorporation and marginalization, even and especially in liberal progressive discourses that make vocal claims to social justice on behalf of marginalized groups while denying their own technologies of power.
>
> *(pp. 6–7)*

When considering gender and educational reform on a deeper level, then, we need to think through the relations of power expressed by these reforms, considering to what extent they are reflective of simple adjustments to superficially include the other or whether these measures truly reflect significant change to practice and social relations by interrupting gender norms. In addition to a thoughtful, feminist reconsideration of critical pedagogy, the same must be done for a host of educational reforms in which gender is prioritized. In the next section of this chapter, we will look at a book containing critical feminist reviews of said reforms specific to mathematics education.

For now, we continue with some specific examples of how a feminist critique can modify our thinking on emancipatory education. One of the authors contributing to *Feminisms and Critical Pedagogy* is Valerie Walkerdine, a feminist psychologist who contributed significantly to gender and education and gender and mathematics education. In her chapter from the Luke and Gore text, "Progressive

Pedagogy and Political Struggle," Walkerdine reflects on her primary school teaching of the 1960s, full of "progressive" intentions to liberate poor, inner-city children from the chains of oppressive economic and state structures. Upon this reflection, she considers the accordance such a project has with *maintaining* the oppressive structures rather than eradicating them. One of her primary examples focuses on the reductive, gendered role of the teacher in such a progressive pedagogy. The teacher, as a political entity, a "bourgeois and nurturant mother" (1998, p. 20), aims to reject any power that might exist in her authoritative presence and, through the freedom she provides, will transform students into rational thinkers. However, the denial of the power that exists within the schooling structure presents a paradox for the teacher whereby students are not free, social-class hierarchies are reproduced, and the teacher is reduced to a feminine, passive subject.

Walkerdine's and other chapters in the anthology also point to the preference for rationalism within critical pedagogy as a masculine priority that reinforces surveillance and controlled societies. This bears significant relevance to the teaching and learning of mathematics, a discipline that can play a big part in categorizing and managing populations. These and the other critiques contained in the volume sit within the third wave of feminism that more carefully draws connections between forms of oppression (social class and gender, for example). This is especially due to the poststructural nature of critique in which the discursive messaging contained in social projects is laid out and scrutinized for their reproduction of particular oppressive structures. As we turn to mathematics education, consider the following exercise. Think about a recent reform in the teaching and learning of mathematics and perhaps specifically one that aims to increase success by female students. How has this reform addressed gender and mathematics, or in what ways is this reform reinforcing gender binaries? The answer must begin by identifying the problems with respect to gender and mathematics, to which we now turn in the next section.

Gender and Mathematics Education

Possibly more so than the other social identities discussed in this book, research on gender and mathematics education is plentiful. Several researchers have synthesized this work into a variety of categories that reflect the varying factions of feminism, and those that we will look at here claim that some of the researchers, more than others, make significant progress toward justice. In this section, I will highlight a few items from this synthesis literature that I encourage you to read and think through. When examining studies that focus on gender and mathematics education, here are a few questions to consider: To what extent does the research view women and girls as lacking something needed to learn mathematics and/or having a pathology that prevents them from doing so? Do the researchers imply a shift in the culture and practice of mathematics education or an adaptation that teachers and caregivers must make to facilitate greater success by girls?

Do the researchers define gender in their study and, if so, how does the definition relate to a poststructural emphasis on the historically constructed nature of gender binaries?

We begin with a synthesis of the literature on gender and mathematics education from Rogers and Kaiser's (1995) edited book *Equity in Mathematics Education: Influences of Feminism and Culture*. In the introductory chapter, the editors lay out a helpful trajectory that summarizes the evolution of research on this topic. Using a framework developed by the feminist Peggy McIntosh, Rogers and Kaiser (1995) present the phases of mathematics education and gender research as follows:

Phase One: Womanless mathematics;
Phase Two: Women in mathematics;
Phase Three: Women as a problem in mathematics;
Phase Four: Women as central to mathematics; and
Phase Five: Mathematics reconstructed (p. 3).

Speaking broadly, the shift in phases starts from the masculine culture and majority-male status of mathematics and ends with changing the nature of mathematics and the culture of its practice. In phase one, a definition of the situation is made clear: Mathematics is a male-dominated space. This resonates with our discussion of mathematics education as a white institutional space, discussed in Chapter 4. Despite the odds against it, some women attain status in the space (as research mathematicians, for example), and reflections shared by these individuals indicate the "silence and exclusion" (Rogers & Kaiser, 1995, p. 4) they experience. In phase two, this problem is addressed superficially with token celebrations of the few women who make it within the male-dominated mathematical world. In response to the problem, the remaining phases make attempts to understand *why* fewer women achieve high status in mathematics. Phase three locates the source of the problem on the women themselves. The phase includes research on and promotion of intervention programs aiming to increase girls' interest and success in mathematics, something that certainly continues to this day and has transformed into STEM education camps and other opportunities for girls. These interventions aim to fix a situation in which girls do not seem interested in mathematics, thereby locating the problem within them. In this case, girls and women are said to have a deficiency. Something they lack is a desire to learn mathematics, unlike their male counterparts. As Rogers and Kaiser point out, another strand of this approach is the study and theorization of math anxiety as a source of the problem. This is an example of an effort to pathologize girls and women; in this case girls and women have an illness that their male counterparts do not. In all cases, these attempts locate the problem within girls and do not work to change the system but work within it (1995, p. 6).

To this point, phase four addresses the problem with a shift toward the system. This research focuses on girls' and women's experiences in learning mathematics and takes a cue from feminist work in psychology, reminding us that how people learn might be influenced by the social identities by which we are positioned. As Rogers and Kaiser put it, phase five suggests that further work will leave us with a new conception of mathematics, and in looking at a broader set of the work in feminist pedagogy and mathematics education, we see that this phase will include pushes beyond the theoretical considerations of phase four as well. In my own view, this active, current phase can include the most advanced conception of gender, via poststructuralist and third wave feminism, to address the intersectionalities of social identity at play as well as consider how conceptions of feminist pedagogy will interrupt the logics of women as inferior Others. This thinking responds to the earlier stances resting on "the ways women think differently," something that can reproduce binary modes of thinking and perhaps perpetuate stereotypes. For the remainder of this section, we will focus in more depth on a few studies from phases four and five. Most importantly, these contributions no longer blame women for their lack of success in mathematics but look to the cultural practices of school mathematics that exclude and perpetuate gender stereotypes.

Recall the earlier section when we discussed the book *Women's Ways of Knowing* (Belenky et al., 1986), which problematized the male-centered theories about how people learn. Rossi Becker (1995) applies this directly to mathematics education and, to begin, provides the following caution in so doing:

> When I refer to women or girls in this chapter, as when the book refers to women in *Women's Ways of Knowing*, I am not automatically referring to all females. Here I am generalizing, not as mathematicians do, meaning every woman, but as social scientists do, meaning most women. The word "women" is used to refer to all those individuals who think, come to know, or react in a fashion that is common to the majority of women. These individuals may be females or males. Also, the use of the word "women" to describe the way some people think does not preclude the possibility that some women do not think in this way. There is, of course, the danger that acknowledging women's different ways of knowing will serve to reinforce stereotypes that demean women's capabilities. We have to make the case that "different" does not mean one way of thinking is better than another. On the other hand, research can help provide evidence to support or refute the possibility of a women's way of knowing in mathematics and, if it is supported, demonstrate how this can help us understand and improve women's participation in mathematics.
>
> *(p. 164)*

These considerations on the limitations of generalizing and potential harm in reinforcing stereotypes are important cautions we must take when thinking about research on gender and mathematics education. In my view, these are more adequately addressed by the research rooted by poststructural feminist thinking, which we will look at later in the section. Nevertheless, the application of *Women's Ways of Knowing* is essential to thinking through gender and mathematics, and we now take a look at some of Rossi Becker's major themes.

Rossi Becker points to *Women's Ways of Knowing*'s suggestion that a "connected teaching" is required to move women toward the higher levels of understanding, such as constructed knowledge. She applies the concept to mathematics:

> In connected mathematics teaching, one would share the process of solving problems with students, not just the finished product or proof. Students need to see all the crumpled papers we put in the wastepaper basket, if they are to understand that mathematicians do not arrive at a solution the first time or the first way.
>
> *(1995, p. 168)*

The suggestions made to foster constructed knowledge resonate with many of the reforms of math education discussed in the second chapter of this book. She emphasizes how we must teach mathematics as a process rather than a "universal truth handed down," that discovery methods should be in favor of traditional modeling and practice, and that multiple methods for problem solving should be encourage in students. All of these are argued to increase success for female students and are consistent with the mathematics education reforms occurring within the trajectory of NCTM standards-based mathematics instruction.

Chapter 4's discussion of social class brought to light an interesting study that pushed back on the reform methods that are promoted by reform mathematics teaching. I bring up Lubienski's study again because it reminds us of the caution that we must take when considering the generalizations presented by *Women's Ways*. To be sure, both the original study and its application aim to study women across social identities (race and class) and yet also articulate a generalized claim that most women think a certain way and will move toward this with a certain form of education. However, Lubienski draws on the literature of reform mathematics education to use whole-class discussions in enhancing mathematical instruction. In studying the girls in the classroom, she found that such efforts (used to promote connected thinking) actually silenced girls from lower-social-class backgrounds. These inconsistencies are exactly the pitfalls that poststructuralist feminist thinking takes on directly. It is important that we think through the multiple positions of individual learners rather than paint with broad strokes and make sweeping modifications that might lead us to consider our job complete. Such efforts that emphasize multiple social identities and reduce binary thinking reflect more advanced contributions in the conversation about gender and mathematics education.

Another chapter from *Equity in Mathematics Education*, an edited book by Australian mathematics education researcher Sue Willis, is titled "Gender Reform Through School Mathematics" (1995). In line with the editors' "phase four," this chapter criticizes research on gender and mathematics that locates deficits or pathologies in girls and supports the research that looks at the system of mathematics education and its specific exclusionary practices. Willis cites research (this is the work of Valerie Walkerdine, a researcher discussed earlier and to whom we will turn again shortly) describing the culture of school mathematics that favors rule following, whereas a challenge of rules and creativity is at the root of superior mathematical talent. This "sets up a conflict between what is regarded as necessary to achieve femininity and be a 'good girl' by doing what the teacher asks, and what is necessary to be regarded as 'really good' at mathematics" (Willis, 1995, p. 191). Similarly, Willis's chapter discusses several projects that implement novel curricula, aiming directly to reduce gender bias and promote greater inclusivity. Interesting, mixed results are discussed, such as the "counter-sexist" mathematics textbook that due to its deliberate nonconformity *and* lack of comment resulted in students and teachers making jokes about the book rather than taking the learning of gender more seriously. As a result, Willis suggests that such intentions to disrupt gender bias need to be made clearer by teachers, authors, and researchers. She also highlights research in which students with privilege in mathematics (typically male students) have openly resisted efforts for greater inclusion. However, had these students known that the efforts were to reduce gender bias in school mathematics, they may have been more receptive to novel interventions.

Moving more into phase five, we look at Valerie Walkerdine's important text *Counting Girls Out* (1998). First published in 1989, the text and its research are positioned firmly within poststructuralist feminism. Walkerdine states her work rests on theoretical developments that put

> into historical perspective the construction of scientific ideas (or truths) about girls and boys, men and women, minds and Mathematics. It allows us to take apart these truths and their forming and informing of practices in which girls and women are taken to be poor at Mathematics.
>
> *(p. 18)*

Further,

> Our starting point is that there is no simple category "woman" which can be revealed by feminist research, but that as feminists we can examine how facts, fictions and fantasies have been constituted and how these have affected the ways in which we have been positioned, understood and led to understand ourselves. Hence, while much feminist counter-research has attempted empirically to disprove the facts about girls' and women's performance, we felt that a fundamental problem remained. Accepting the categories and terms within which the issues were framed left feminist work always on the defensive and trapped within empiricism.
>
> *(p. 21)*

In the book, Walkerdine describes the fundamental character to school mathematics as the teaching of reason and the supremacy of rationalism. In her efforts to think of mathematical culture as historically embedded, she traces the roots of Cartesian rationalism to the enlightenment and highlights its gendered, specifically male-dominated, nature. Women were excluded and deemed irrational, their emotional characters and bodies only in service to the reproduction of reasoned man. What students learn is not their facility with mathematics but their capacity to reason, to break free of emotion. This is the culture, practice, and ethos of the space of mathematics education. In addition, there exist steadfast gender binaries in our society, habits of mind that dictate who we are and how we should behave. Given that teachers and students both are subjected to constant messaging about these expected gender roles, it comes as no surprise, then, that girls and women are seen as outsiders to the space of mathematics education.

Walkerdine's extensive study of the mathematics education space supports this. The book recounts several studies across grade levels. One research project interviewed students and teachers to come to understand just what makes a student labeled "good" or "bad" in school mathematics. Walkerdine and her research team found stark contrasts. The following excerpt provides an example of this:

> Angela, then, is positioned classically as a "good girl," an "ideal pupil," but she does not have that elusive gift, "brilliance." She must rely on hard-work. But what has happened here? We saw how "sensitive" Angela's mother was when the child was 4. She is articulate, one of the few children who found it easy to chat to her nursery school teachers. Why then at 10, even though she is outstanding in her class (surely coming top is an indication of achievement), is she designated a quiet, shy girl who comes top only through sheer hard-work? This judgment was never applied to any boy in any study—on the contrary, it was difficult for a boy to be judged a failure, even with the most appalling performance. Teachers talk, for example, about boys with very poor attainment still having potential or being bright.
>
> *(p. 88)*

The story of Angela is not unique and represents the consistent efforts Walkerdine and her research team took to inquire and discuss the school math experiences for girls. In study after study, Walkerdine finds teacher talk to indicate that boys who do well are gifted yet girls who do well work hard.

Since Walkerdine's seminal text, two books have continued this trajectory of poststructural feminist inquiry into mathematics education. *Masculinities in Mathematics* by Heather Mendick (2006) continues the work of thinking through what characterizes good and bad math students. Subjects in her studies demonstrate that the gendering of school mathematics is just as much about social norms and expectations for work life as it is about "the gendering of maths itself, the gendered phantasies of rationality and of genius, and the ways these work with the construction of masculine and feminine as oppositional categories" (p. 67). A text clearly situated within third-wave feminism, Mendick discusses the "queering of mathematics education" as well. Along the lines of a poststructural refusal of binaries, one example of such efforts requires math teachers and students to reclaim the notion of ability. Throughout her research, exceptional mathematics students are thought to be born that way, yet we now know that who is really perceived as and encouraged to be rational falls along race, class, gender, sexuality, and other lines. A consistent rejection of the inborn nature of ability in mathematics education is an excellent suggestion for mathematics teachers to consider.

The second book, by Sara Hottinger (2016), also carries on the tradition of poststructural feminist investigations of mathematics education. *Inventing the Mathematics: Gender, Race, and Our Cultural Understanding of Mathematics* requires that "We as a culture need to challenge, in a very deep and complex way, how we construct what it means to reason, what it means to think logically, and what it means to think mathematically" (p. 24). Hottinger provides exceptionally interesting poststructural analyses of mathematics textbooks, the field of ethnomathematics, and even the actress-turned-female mathematician Danica McKellar. All of these endeavors relate to the objective of calling into question what and who mathematicians are and, as the title suggests, motivate a new conception for them.

This section has focused more on the research in gender and mathematics education that pushes our thinking away from gender binaries as fixed realities. Other efforts are not considered to be critical because they ultimately reinforce gender binaries and stereotypes. The research that questions why female students do not like math usually pathologizes or otherwise blames girls. Efforts such as addressing female students' math anxiety reinforce gender binaries with further attempts to subtract feminine emotions. To encourage mathematics education that is more inclusive for women, we must look deeply at the masculine culture of school mathematics as well as the gender binaries at play in larger society. In reviewing these contributions made here, we might consider gender and mathematics education to have pushed the notion of social construct more fully than the other social identities discussed in this book. As well, the poststructural feminist work on mathematics education more fully interrelates the competing social identities of race, class, and gender in the spirit of the third wave. Teaching mathematics critically requires such an intersectional perspective, and we will explore this idea more fully in the final chapter as we "put it all together."

Activities and Prompts for Your Consideration

1 Much of the less critical work on girls and mathematics education has shifted to girls and STEM education. Locate a project (for example, an after-school program, a summer camp, or perhaps a research study) and read critically the language used to introduce the program. How does the STEM program frame the problem with girls and STEM? Does the organization and/or its

program consider other social identities besides gender, such as social class or race? What are other ways that a poststructural feminist position can critique the project?

2 Consider Mendick's (2006) notion of "queering mathematics education" that calls into question the all-too-common belief that you are either born good at math or not. Choose one or a few of the following activities:

   (a) Write your own mathematics education autobiography, focusing specifically on moments when you felt "you had it" or "you didn't have it." How did certain people in your life reinforce or disrupt gender binaries during those times? For examples of mathematical autobiographies, take a look at Hottinger (2016) and Mendick (2006), as well as Hottinger's brief biography of Danica McKellar.
   (b) Write out a "to do" and "not to do" practical guide for mathematics teachers who want to disrupt the notion of ability.

3 Our work in teaching mathematics with a critical perspective sometimes requires that we attempt to get others to think differently as well. Would you engage the following teacher if he came to you as described in what follows? If so, what would you say? How does your own positionality (age, race, class, gender, sexuality, ability) affect the actions you might take in this hypothetical situation?

At the end of one school day, Bill, a younger white teacher across the hall and whom you've known for about 3 months comes to you to discuss his "project student," the one he seems to like talking about to you. When he typically talks about June, a Black girl who consistently struggles in his Algebra 1 class, he likes to point to her hard work that gets her to perform fairly well on the tests. However, today Bill is less optimistic. He comes the day that they reviewed for their first comprehensive exam, and he was so saddened to see that June could not remember many of the skills that she had mastered just months and, in some cases, weeks prior. Bill questioned whether June's hard work, and his, was even worth it. He was starting to think that mathematics learning was something that just wasn't in the cards for June.

References

Belenky, M. F., Clinchy, B. M., Goldberger, N. R. & Tarule, J. M. (1986). *Women's ways of knowing: The development of self, voice, and mind.* New York: Basic Books.
Butler, J. (2004). *Undoing gender.* New York: Routledge.
Foucault, M. (1990). *The history of sexuality: An introduction, Volume 1.* New York: Vintage Books.
hooks, b. (2014). *Feminism is for everybody: Passionate politics.* New York: Routledge.
Hottinger, S. (2016). *Inventing the mathematicians: Gender, race, and our cultural understanding of mathematics.* Albany, NY: SUNY Press.

Luke, C. & Gore, J. (1992). *Feminisms and critical pedagogy*. New York: Routledge.
Mendick, H. (2006). *Masculinities in mathematics*. New York: Open University Press.
Rogers, P. & Kaiser, G. (1995). *Equity in mathematics education: Influence of feminism and culture*. Washington, DC: Falmer Press.
Rossi Becker, J. (1995). Women's ways of knowing mathematics. In *Equity in mathematics education: Influence of feminism and culture*, edited by P. Rogers & G. Kaiser. Washington, DC: Falmer Press, pp. 163–174.
Walkerdine, V. (1998). *Counting girls out: Girls and mathematics*. Bristol, PA: Falmer Press.

6

PUTTING IT ALL TOGETHER

Intersectionality, Current Mathematics Education Policy, and Further Avenues for Exploration

Each of the last three chapters has taken on a prominent social identity by thoroughly discussing the relevant literature both within and outside of research on mathematics education. This aims to promote a critical practice in mathematics education that engenders inclusion in both content and its access to it. Recall that in Chapter 2 we also dealt similarly with the social identities of language-minority students and students with disabilities. Here and by way of conclusion, we will put these isolated discussions together and in further action in three ways. First, we will break down the false implications of such isolated discussions of social identity and instead favor the notion of intersectionality, in which mathematics students as learners live in fluid spaces of contesting identities. In other words, the previous chapters and discussions should not force us to view any student as simply "not white" or "working class" or "female"; we should instead see that each individual and all communities act within and through an intersection of multiple social identities constructed by sociopolitical formations of relations between social groups. This discussion falls directly out of our previous discussion of third-wave and Black feminism of the preceding chapter. Second, it is important to situate our work of critically teaching mathematics within the current context. To do so, I provide a brief review of the history and politics of mathematics education, a somewhat hard to swallow tale for critical mathematics teaching. Following this, the third section responds with a review of essential readings for critical mathematics educators. The previous chapters have covered these as they relate to race, class, and gender, but here I put together a sketch of specific anthologies and journals for further reading. This balances out the second section by providing hope in the midst of dark policy times. Put together, the three sections of this chapter aim to properly situate the preceding chapters in a more comprehensive and total framework.

Although at times these topics will complicate and make more difficult our work, they are necessary discussions that make our objectives more realistic, practical, and steadfast.

Intersectionality

The danger in our previous discussions is that we might walk away thinking of social identities in isolation from each other or perhaps position one over the other as we work to teach mathematics critically. Vivian May refers to this as a single-axis world in which too often one of the social identities is privileged over another. For example, there are several examples of Marxist scholars who think that economics can explain everything, that it is the first social identity that explains race and gender. And for every type of Marxist scholar of this accord, we can find a race scholar who similarly points to race as a "first identity." Consider another framing, however, in which social identities are framed as interwoven and notions of power as complex and dynamic.

Such thinking, what we look at in depth here, has been termed intersectionality. It promises to carry forth the traditions set up by third-wave and Black feminists. Recall for example the contributions of bell hooks that we discussed in the previous chapter. Her accounts and theorizations of feminists within both the civil rights movement and the feminist movement fell short of describing and accounting for the Black female experience. Worse, in failing to acknowledge the multiplicity of oppression, they both in their own ways served to reinforce oppressive structures for the causes that they did not identify.

May (2015) describes intersectionality as follows:

> First, it approaches lived identities as interlaced and systems of oppression as enmeshed and mutually reinforcing: one aspect of identity and/or form of inequality is not treated as separable or as superordinate. This "matrix" worldview contests "single-axis" forms of thinking about subjectivity and power and rejects hierarchies of identity or oppression. An intersectional justice orientation is thus wide in scope and inclusive: it repudiates additive notions of identity, assimilations models of civil rights, and one-dimensional views of power . . . For instance, its matrix model changes the terms of what "counts" as a gender, race, sexuality, disability, nation, and/or class issue or framework. Intersectionality also approaches lived identities, and systemic patterns of asymmetrical life opportunities and harms, from their interstices, from the nodal points where they hinge or touch.
>
> *(p. 3)*

A critical approach to teaching mathematics, then, would not prioritize racial equity or gender equity, for example. It also would not treat these issues in isolation from one another.

Instead,

> By focusing on how patterns and logics interact, and how systems of oppression interrelate, intersectionality highlights various ways in which, unwittingly, we can be engaged in upholding the very forms of coercion or domination we seek to dismantle. It is thus indispensable for identifying paradoxical outcomes and for revealing unexpected or hidden points of contact between the liberatory and the coercive. Via its matrix orientation, and attention to relational power and privilege, simultaneity, and underlying shared logics, intersectionality needs to be understood to have explanatory power, analytical capacity, and normative political component, one focused on eradicating inequality and exploitation.
>
> *(May, 2015, p. 5)*

Recall the research in mathematics education that focuses only on reform mathematics as a means toward equity and the ways that this has reinforced privilege by not acknowledging the culture and linguistic diversity of learners. Instead, the focus of research that informs our practice must look for the matrix of identity and toward the goals of resisting hierarchies and domination as they exist in multiplicity.

Mathematics education scholarship has begun to reveal the natures of such intersections. Riegle-Crumb and Humphries (2012) discuss the ways that gendered discussions of math performance in which males outperform females contain hidden racial connotations as well: "Stereotypes of men's innately higher math ability refer specifically to white males" (p. 295). Their research provides further study of the complicated nature of teacher perception of ability, in which the multiple social identities that learners occupy is reflected in a teacher's perception of his or her ability. These are the types of discussions that the lens of intersectionality affords and that mathematics educators can reflect on. Holding a single-axis view, as so many researchers have done, fails to acknowledge both the systemic nature to oppression across social identities and the multiple oppressive structures that learners will face within the school structure.

In tracing back intersectionality to the Black and third-wave feminism, one of its essential features is the consistent application of hierarchy and domination that pervades western industrial culture. The goal, as May puts it, is the eradication of exploitation in all its forms. Although somewhat distanced from intersectionality, another strand of feminism points to this consistency and ties in all forms of exploitation and hierarchy. Ecofeminist philosophy targets the systems of oppression systemic to modern times including male superiority, white supremacy, social-class hierarchy, ableism, and human supremacy. As an example, ecofeminist Karen Warren (2000) puts it as follows: "The dominations of women, other human Others, and nonhuman nature are interconnected, are wrong, and ought to be eliminated" (Warren, 2000, p. 155). In this vein, ecofeminists add human supremacy to the list of exploitative structures.

Strands of ecocritical scholarship in education are emerging throughout the globe and with great popularity, especially in the field of curriculum studies. This has yet to be applied to mathematics education, but the critical orientations toward mathematics education scholarship will do well to explore these in the coming years. For now, explorations are laid out nicely that apply ecofeminist and other ecological thinking to education, in Martusewicz, Edmundson, and Lupinacci (2015). Some of the key points that mathematics education can consider are EcoJustice Education's commitments to exposing and "unlearning" the discourses of modernity, or habits of mind that pervade our thought and align with the logic of domination. As with a focus on intersectionality, the goal of EcoJustice Education is the eradication of exploitation and hierarchy.

Bringing intersectionality and ecofeminism to mathematics education would aim to eradicate exploitation by providing all people with access to mathematics as well as altering the content of mathematics instruction. Mathematics teaching and learning conceived in this manner would bring together the best ideas from sustainability mathematics education and social justice mathematics education, without prioritizing one over the other. This is not the same as equating, for example, the oppression of nonwhite persons to the exploitation of animals. Rather, it is unlearning the dominating habits of mind that pervade our culture and reconstituting assumptions and frameworks that support interspecies mutual aid and just communities. Mathematics teachers with critical orientations can look for the ways to insert such discussions into their mathematics teaching; at the very least, they commit to practices and reflection on their practice that do not dominate or exploit learners as the hierarchies of social identities present in the modern world typically would have them do. At this time, there are more examples of mathematics education units that prioritize one aspect of exploitation than ones that allow students to make these connections. I expect that poststructural, intersectional, and ecofeminist applications to mathematics education will occur in the coming years to advance our work in teaching mathematics critically.

In similar ways that draw on differing traditions, the theories of intersectionality and ecofeminism bring together work that examines exploitation and domination in isolated instances. Teaching mathematics with a critical perspective would do well to pay attention to these rather than commit to liberatory practices that might serve one aspect of domination while perpetuating another. It would do well to follow in the traditions now set forth by mathematics education scholars, such as Gutstein (2009) in Chapter 4, Larnell, Bullock, and Jett (2016), mentioned in Chapter 3, and Walkerdine (1998) in Chapter 5, who position their work on race and mathematics education within an intersectional framing.

History and Politics of Mathematics Education

Our next discussion that motivates teaching mathematics with a critical orientation comes with a review of the history and politics of mathematics education. The contents of this section piece together from a variety of sources, including mainstream histories of mathematics education within the research community as well as critical perspectives on mathematics education, including my own work in Wolfmeyer (2014) (and the work of one conservative mathematics educator, Klein, 2003) as well as discussions of history and politics of education in general, such as Joel Spring (2000). Simply put, mathematics education is not exactly a friendly space to the critical perspectives discussed throughout this book. Reviewing the history reveals that mathematics education policy directly conflicts with critical goals; as we shall see, military and business interests have been the primary motivators for the development of mathematics education. This history and politics will be focused on the United States; however, similar trends exist across the globe, and we will look at how these have spread through the increasing use of standardized tests.

This history and politics review begins with an assertion of a traditional mathematics education during World War II, followed by the new math of the Cold War, yet another traditional, back-to-basics movement in the 1970s, a new-new math movement of the 1990s, the insertion of the standardized testing industry under No Child Left Behind, and finally the development and implementation of a national mathematics curriculum called the Common Core. As well, we look at the influence of two international mathematics tests, the Trends in International Math and Science Study (TIMSS) and the Programme for International Student Assessment (PISA). In tracing the histories through a focus on politics and policy, we will make clear a battle over traditional versus reform pedagogy and the primary business and military interests at play in mathematics education.

To begin, first consider the educational stage set by reformers of the 1920s and 1930s. It was commonly referred to as the progressive era, and several aspects were at play. On the one hand, the progressives aimed to make an efficient, orderly education that streamlined the process and placed people according to the roles society had determined for them. On the other, educational experiments built off of the philosophies of John Dewey, who asserted that education be rooted in experience. In this framing, mathematics was taught through experiences and out of necessity. For example, students might work on a carpentry project and thus be required to learn fractions.

With this progressive era as the backdrop, military leaders during World War II claimed their soldiers had poor training in basic mathematical skills. The most famous of these, Admiral Nimitz, declared that more specific and direct instruction in both basic and advanced mathematics was needed for naval officers. Such a motivation for a strong U.S. military was well received by the populace given the fight in a so-called good war to defeat fascist Europe. Nevertheless, this first national conversation clearly situates a national mathematics education within the needs of the U.S. military.

Continuing these military motives, the 1950s Cold War era pushed national mathematics away from traditional, basic skills emphasis and toward what we now call reform mathematics teaching. Although this may seem a surprising about-face, the concerns about *Sputnik* and the Soviet Union's outcompeting the United States in the arms race motivated the U.S. to develop a curriculum that groomed the nation's "best and the brightest." This translated to mathematics curriculum that taught students early on to think abstractly like mathematicians do, to set them on a path toward advanced mathematics, engineering and physics. A team of mathematicians, headed by Ed Begle, developed a "new math" curriculum for use in schools. National curriculum is not possible in the U.S., due to the separation of powers among federal and state and local governments. However, the federal government was able to have de facto control over the curriculum by providing funds to states and local education agencies that promised to use this new curriculum developed for the military interest (Spring, 2000). Thus, new math found a home in public schools throughout the country.

The 1970s saw a backlash movement against the new math. This was comprised of teachers and parents but spearheaded by mathematicians and professors, who were not pleased with the preparation of students arriving in their introductory college classes.

> In 1962, a letter entitled *On The Mathematics Curriculum Of The High School*, signed by 64 prominent mathematicians, was published in the *American Mathematical Monthly* and *The Mathematics Teacher*. The letter criticized New Math and offered some general guidelines and principles for future curricula.
>
> *(Klein, 2003, p. 185).*

Thus the 1960s new math and 1970s back-to-basics movement launched an ongoing debate among mathematicians and applied mathematicians commonly known as the math wars. On the one hand, you have the traditionalists who emphasize fact mastery and skills, with the logic that students can learn these first and then understand the concepts and theory of mathematics later. On the other hand, the reformists emphasize that mathematics education should mimic the work of mathematicians, with a focus on the process of mathematical thinking. At the policy level, thus far the debate over national mathematics curriculum was motivated by military defense and security.

Moving to the next decade, the 1980s Reagan era shifted this focus. With the publication of the policy brief *A Nation at Risk*, the goals of mathematics education moved toward economic security and dominance. This document detailed the threat against the United States economy by the economies of West Germany and Japan. With clear commitments to mathematics and science, these efforts called for national curriculum that would better prepare students for a workforce to compete in the global economy. Continuing in the trajectory set up by new math, the reformists, mostly research mathematics educators, answered this call for a national mathematics curriculum in 1989 with the National Council of Teachers of Mathematics (NCTM) *Standards*. These standards are framed by economic concerns and more closely orient toward a reform mathematics program, emphasizing process over skills and fact mastery. Their updated standards in 2001 somewhat compromise this stance but still remain entrenched in what has been termed a new-new math. Yet again, these standards were not officially a national curriculum. However, the National Science Foundation supported research and development of standards-aligned curriculum that ultimately found its way into a majority of textbooks used in public classrooms.

Also in 2001, the U.S. Congress passed the reauthorization of the Elementary and Secondary Education Act (ESEA) with the title No Child Left Behind (NCLB). As with each reiteration of ESEA since its origins in the 1960s, NCLB prioritized a fair and just education for all students regardless of gender, socioeconomic status, or race/ethnicity. NCLB's take on this was the requirement that all states have standards and standardized assessments in place. Again, the federal government does not have control over education but has implemented such national education policy by tying requirements to a state's receipt of federal funding. NCLB clearly situates within concerns over preparing a strong workforce; a lesser-known fact of NCLB is that local districts are required to provide the names and contact information of students to military recruitment officials. Therefore, NCLB continues the theme of commitments to the U.S. economy and military.

Perhaps more impactful on mathematics education, NCLB's policy requirement of standardized testing introduced a host of educational businesses (nonprofit and for profit) that would provide standardized testing services. These assessment companies were interested in a unified national curriculum for the sake of efficiency; delivering products to 50 different states was more costly than delivering one product. In addition, the math wars continued to battle over the new-new math's overemphasis on process and looked to balance the curriculum with a traditional focus.

Along the lines of a global economic competition, there has existed a global competition among the world's mathematics students. Nations compete against each other for the top math scores among their general student populations. Primarily two tests are viewed in this view: the Trends in International Mathematics and Science Study (TIMSS) and the Programme for International Student Assessment (PISA). Both have strong commitments to educating for the economy; TIMSS was at one point sponsored by the World Bank, and PISA is sponsored by the Organisation for Economic Cooperation and Development (OECD). Putting these points together, performance in mathematics education is seen primarily as an indicator of economic possibility. Ever more and more countries are added to the list of participation in the TIMSS study that parallels the spread of capitalism throughout the globe.

Back to the United States, mathematics educators and mathematicians looked to other countries that outperformed the U.S. on these tests, as it turns out, to justify a more traditional curriculum. Using phrases "internationally benchmarked" and "coherent curriculum," U.S. researchers looked specifically to the national curriculum programs of authoritarian capitalist countries like Singapore and South Korea. These curricula emphasize a highly sequential structure to learning mathematics that the NCTM new-new math standards lacked. Ultimately, these research efforts led to the development of the Common Core State Standards for Mathematics (CCSSM). Also involved were the standardized testing companies so eager in promoting a national curriculum for their profit. And, as usual, these standards were made a de facto national curriculum through the Race to the Top program of 2011, in which states competed for federal funds and were required to adopt the CCSSM in order to do so.

My own (Wolfmeyer, 2014) study on how CCSSM came to be highlights in great detail the extent to which the economic priorities are in place, as well as the testing and other industries' influence on mathematics education. The majority of individuals and organizations involved in writing national mathematics education policy hold firmly stated commitments to a public education that increases corporate profit. As Joel Spring (2000) writes, there is an inherent tension between a public education for the people and a public education for the development of corporate profit. For example, the 1980s fostered a greater relationship between schools and corporate profit, a time when we also witnessed a greater stratification of wealth.

Other industries at play in mathematics education include the information and communications industry, mostly because standardized tests can now be delivered in digital format. The influences on mathematics education are now not only interested in developing these assessments in digital format but in using digital instruction as well. Major players stand to gain great profits through the activity of public mathematics education, including companies like Microsoft, Apple, Pearson, and ETS. Finally, my analysis of mathematics education policy reveals a continued interest in national mathematics education for the U.S. military.

Such a history of mathematics education policy in the United States and its spread across the globe indicates strong commitments to educating for corporate profit and for strengthening the military. These goals are counter to education with a critical perspective that would seek to promote democratic goals and personal fulfillment. Certainly such a context makes our work in teaching mathematics critically more difficult, but it is highly important to recognize these challenges and work toward change. In this vein, the next section's selection of critical points of interest from mathematics education should provide hope and a sense of solidarity in your work.

Searching for Allies: Further Avenues for Exploring Critical Work in Mathematics Education

The majority of resources drawn on within this book, especially in Chapters 3 through 5, come from the more critical strands of mathematics education. In my efforts to familiarize you with the landscape of critical perspectives on mathematics teaching, I conclude this book with introductions to some further reading and areas for exploration along these lines. As it turns out, many mathematics educators identify with the label critical mathematics education (CME), and some do this kind of work but might not accept the label. My goal here is to sketch a landscape of this work as a response to the current context described in the previous section. I encourage you to seek out these spaces for further exploration and thinking about mathematics teaching and learning. We now review one anthology not yet discussed as well as the journals in which you can typically find critical work in mathematics education.

One recent edited collection on critical work is *Opening the Cage: Critique and Politics of Mathematics Education* (2012) by Ole Skovsmose and Brian Greer. It contains chapter essays written by some of the most prominent mathematics educators arguing from a critical perspective. Its introduction, written by Skovsmose and Greer, traces the origins of what they term a "critical mathematics education" and reference several key scholars and concepts. One of these contributions is Skovsmose's work himself: an almost 40-year-long quest to theorize a critical mathematics education, which he began in the 1980s and continues today. Another topic Greer and Skovsmose point to is the destabilization of mathematics as exclusively western, particularly in the work of D'Ambrosio's launch of the field of ethnomathematics. A third strand of critical mathematics education pointed to in the introduction is the specific application of Paulo Freire's praxis to mathematics education, initiated by Marilyn Frankenstein and further developed by Eric Gutstein.

Speaking more broadly about the more critical work in mathematics education, Skovsmose and Greer write:

> The field of mathematics education, in general, has considerably matured, as reflected in the diversification of influential disciplines and related methodologies—broadly speaking, the balancing of technical disciplines by human disciplines such as sociology, sociolinguistics, anthropology, psychoanalysis, and of formal statistical methods by interpretative methods of research and analysis. Within the field, there is heightened cultural and historical awareness, both within and beyond academic mathematics, and an increased acknowledgment of the ubiquity and importance of "mathematics in action" and the implications for mathematics education, including more curricular prominence for probability, data handling,

> modeling, and applications. In relation to the political nature of the enterprise, there is greater attention to the relationships between knowledge, education, and power.
>
> *(2012, pp. 3–4)*

The introductions to thinking about mathematics education critically that have occurred throughout this book are the results of this turn in mathematics education research.

Similar to this chapter's preceding section, Greer and Skovsmose paint a sad picture of present-day mathematics education: "A great deal of the world's intellectual talent in mathematics (and science) is used in the creation of better ways of killing, subjugating, or surveilling and controlling people, of which current deployment of flightless aircraft, 'drones,' provides a chilling example" (Skovsmose & Greer, 2012, p. 5). They describe mathematics as cast in "an illusion of certainty" by which

> people abdicate the responsibility of making judgements in complex social situations. [As well,] people and institutions within mainstream mathematics education too often collude with the political establishment by willfully remaining oblivious of the social and political contexts outside their self-constructed cage.
>
> *(Skovsmose & Greer, 2012, p. 5)*

The edited anthology contains several and varied responses to such a context of mathematics, mathematics education and the world. One of these is by Eric Gutstein, in which he announces that "mathematics is a weapon in the struggle." By this he suggests that mathematics should not be discounted for its relationship to power and destruction, as mentioned, and instead harnessed for its potential to read and write the world. In the chapter, Gutstein documents his research efforts in teaching an urban mathematics classroom, where he taught discrete dynamical systems through a unit on HIV/AIDS in the students' community. The students modeled the epidemic and through this negotiated several difficult mathematical concepts and skills as well as challenging social, economic, and political conversations. Moving far beyond the mathematics curriculum that urban students typically learn, the unit also engaged with social theory concepts, such as the intersectionality of social identities (race, class, and gender) that many do not learn until graduate school.

Other essays in the anthology include Alexandre Pais's critique and deep theoretical exploration of the strand in mathematics education research that focuses on equity, arguing that such research reinforces exclusionary practices rather than achieving the goals its rhetoric implies. Marta Civil's work on a critical mathematical education for immigrant students is also included in the anthology, as well as Sikunder Ali Baber's discussion of mathematical education in Pakistan and other international perspectives. In total, the chapters in *Opening the Cage* move between critique and possibility, providing a sure understanding of the context in which critical teachers of mathematics find themselves but also the potentials and inspirations toward which we can strive.

As the authors point out, the world of mathematics education research is large and full of variations. Many individuals and organizations commit to an objective of mathematics education aligned to corporate profit and war, at the very least by not taking direct stances against this. Others take steps toward "equity" but do not embrace fully the social theory explained throughout this book, from understanding white supremacy to critiques of capitalist logics. In other words, the field of mathematics education can be daunting for a novice critical teacher of mathematics, and it is important that allies identify the spaces in which critical work is published and communities where the work is promoted. These final, subsequent paragraphs point you to the academic journals and communities in which you will find critical orientations to mathematics.

Two international research mathematics education journals do not have explicit orientations to critical work in mathematics but have fostered a space for this work within their objectives. *For the Learning of Mathematics* and *Educational Studies in Mathematics* contain a variety of topics relevant to research in mathematics education, but authors such as those throughout this book regularly publish in these journals. Although browsing these journals will demonstrate that some research fails to acknowledge the social and political dimensions to mathematics education, at least a few articles in most issues will do this. Two journals in the United States have more critical orientations and reputations. These are *The Mathematical Enthusiast* (formerly *The Montana Mathematical Enthusiast*) and *Journal of Urban Mathematics Education*, created recently by David Stinson and now edited by Erika Bullock. This is not to suggest that other mainstream journals like the *Journal for Research in Mathematics Education* do not contain critical work. However, as an example, this journal prioritizes articles that contain empirical findings and is less likely to publish articles that push the theoretical boundaries of critical mathematics.

Finally, with some digging and searching, you will find that work in critical mathematics finds a home elsewhere. At the risk of self-promotion (which I am also guilty of elsewhere in this book), this has been the case for some of my own writings. In a foundations of education journal from the United States, *Educational Studies*, you will find my article "In Defense of Mathematics and Its Place in Anarchist Education" in which I use anarchist theory as a lens through which to critique the politics of mathematics education as well as sketch an alternative. This journal routinely publishes educational perspectives across the disciplines but mostly those with critical orientations. A second example, I wrote the article "'Math for America' Isn't" that critiques the organization that suggests mathematics teachers be paid more than other teachers. The critique rests on the organizations implications for the value of education as well as its affiliations to corporate profit. This article found a home in the *Journal of Critical Education Policy Studies* because it likens itself to the policy analyses contained in this journal rather than the types of articles in the preceding journals. All of this is to suggest that whether you are a reader thirsty for more critical mathematics work or an emerging writer looking for publication venues, you should consider the framing of journals carefully and search in both mathematics education journals and critical education journals.

In this concluding chapter, I have attempted to bring the previous, isolated discussions on mathematics, teaching mathematics, race, class, and gender together by looking at intersectionality, the history and politics of mathematics education and further avenues for exploring critically teaching mathematics. Intersectionality serves to make our work more complex with a matrix approach to understanding individuals, communities, institutions, and power. In teaching mathematics critically, such an intersectional view privileges the pervasive logics of exploitation and domination that occur throughout western industrial culture. A critical look at the history and politics of mathematics education reveals just how much we are up against yet also is essential knowledge for us to determine where spaces exist to insert our critical orientations. Finally, I have pointed you to some spaces that embrace critical orientations to and critiques of mathematics education, including academic journals that I encourage you to browse and select articles to read.

Thus my final contribution, and indeed the spirit throughout this book, has been calling you to action. Some readers may feel sympathetic to certain arguments contained in the book and others less so. My goal has been to expose you to the more critical perspectives on mathematics education such that you can form your own opinion. No matter your stance, after reading this book, I doubt you can honestly make the claim that mathematics education is objective and agreed upon by all. Mainstream mathematics education might not state its assumptions of corporate power and war or its orientation as a "white institutional space" or its masculine obsession with rationalism, but these political, economic, and thereby oppressive components exist in it. Critical mathematics educators recognize this and critique, as well as propose and practice alternatives.

And, for those of you that embrace most or many of the critical orientations included throughout, I encourage you to take up this world of teaching mathematics critically in both practice and theory. We need active minds working in communities who educate learners in mathematics. They should be literate in the debates and controversies on mathematics education rather than complicit with what is handed to them. As well, we need more minds contributing to the dialogue. Critical orientations to mathematics education have been worked on for at least 40 years, and as such, this is a young discourse. The essential spirit of this community is to challenge, to dig deeper, to question. As Skovsmose and

Greer (2012) remind us, such critique in this community has at once a notion of challenge *as well as* today's crises at heart. Please, challenge mathematics education for the sake of justice, for peace, for all living things having more control over decisions that impact them, and for an inhabitable planet with all life at its center rather than humans. I ask you to, in this spirit, challenge the work in mainstream mathematics education *and also* to challenge the words I have written in this book and the many who engage in teaching mathematics from a critical perspective.

Activities and Prompts for Your Consideration

1 Pick an aspect from reform mathematics teaching, such as the use of whole-class discussion, that we covered in Chapter 2. Think critically and with an intersectional framework about how individuals and groups of students with particular, multiple social identities will approach the teaching method as suggested by reform mathematics. How can the teaching method be adapted for a more inclusive mathematics education?
2 After reading the section on history and politics of mathematics education, what do you think about a national curriculum for mathematics? Should one exist and, if so, what should its goals and orientations be?
3 Browse articles in the journals mentioned in the third section. Select an article by an author that is new to you and share with your peers.
4 Where do you stand in terms of critically teaching mathematics? Do you agree with some but not all of the critiques contained throughout this book? How do you think differently about mathematics education now that you have read this book?

References

Klein, D. (2003). A brief history of American K–12 mathematics education in the 20th century. In *Mathematical Cognition*, edited by James Royer. Charlotte, NC: Information Age Publishing, pp. 175–259.

Larnell, G., Bullock, E., & Jett, C. (2016). Rethinking teaching and learning mathematics for social justice from a critical race perspective. *Journal of Education 196* (1): 19–30.

Martusewicz, R., Edmundson, J. & Lupinacci, J. (2015). *EcoJustice education: Toward diverse, democratic and sustainable communities* (2nd Edition). New York: Routledge.

May, V. M. (2015). *Pursuing intersectionality, unsettling dominant imaginaries*. New York: Routledge.

Riegle-Crumb, C. & Humphries, M. (2012). Exploring bias in math teachers' perceptions of students' ability by gender and race/ethnicity. *Gender and Society 26* (2): 290–322.

Skovsmose, O. & Greer, B. (Eds.) (2012). *Opening the cage: Critique and politics of mathematics education*. Boston, MA: Sense.

Spring, J. (2000). *The American school: 1642–2000* (5th edition). New York: McGraw-Hill.

Walkerdine, V. (1998). *Counting girls out: Girls and mathematics.* Bristol, PA: Falmer Press.
Warren, K. (2000). *Ecofeminist philosophy: A Western perspective on what it is and why it matters.* Lanham, MD: Rowman & Littlefield Publishers, Inc.
Wolfmeyer, M. (2014). *Mathematics education for America? Big business, policy networks, and pedagogy wars.* New York: Routledge.

GLOSSARY OF TERMS

Absolutism A branch of philosophies of mathematics committing to mathematics as a fixed, indisputable truth. Includes logicism, in which mathematics is reduced to logical terms and principles; intuitionism, essentializing mathematics as the creation of objects and proofs about them; and formalism, as a mathematics that defines and works in symbols. These contradict fallibilist and social constructivist philosophies of mathematics.

Achievement gap The conversation regarding the differentials in math test scores between white and nonwhite students, among other social groups as well. Stated to motivate our attention toward better education for all, but has been argued to perpetuate myths about students of color by calling attention to their inferior performance on tests, which can often be racially biased to begin with.

Allyship When a member of a dominant or majority group rejects the privileged status of said group and actively works toward ending the oppression. Implies that allies do not lead movements but support them in ways that are appropriate, such as educating fellow members of the dominant group.

Applied mathematics A major field in mathematics that makes sense of the physical and social world through the application of mathematical concepts.

Closure The ending to a mathematics lesson that provides a clear summary of what was learned during the lesson, an opportunity for the teacher to check for understanding, and setting the stage for how the material will be used individually or in the next group lesson.

Common Core State Standards for Mathematics (CCSSM) In the U.S., a de facto national mathematics curriculum that represents a compromise between reform and traditional mathematics education. "Internationally benchmarked," meaning influenced by the national mathematics education standards of authoritarian capitalist countries.

Critical pedagogy A tradition begun by Brazilian educator Paulo Freire and drawing from Marxism. Suggests an education by which oppressed people can raise their consciousness of their own experience, particularly the social relations and power between groups of people.

Critical race theory Applied to education, critical race theory questions the interrelation among race, power, and inequity and begins by looking at unequal access to quality schools.

Cultural capital As the knowledge, behavior, and habits of mind for success in the economic order, this is what is taught and assessed by schools. Upper-class students have advantages because they learn it both in and out of school.

Culturally relevant pedagogy As a method of teaching nonwhite students, culturally relevant pedagogy has three facets: academic excellence, cultural competence in one's own culture, and a critical consciousness in coming to know the world and one's place in it.

Disability studies The critical view on disability suggesting it as one among many aspects to human difference. As a social construct, disability engages with social, political, and economic notions of power and, as such, an individual labeled disabled is socially perceived as inferior, similar to other social constructs like race, class, and gender.

Ecofeminism A branch of feminism that highlights an underlying logic of domination throughout western industrial culture, via such hierarchized dualisms as male superiority, white supremacy, social class hierarchy, and human supremacy.

Emergent bilinguals For denoting language-minority students, a more accurate term than "English language learners" that critical educators use. Suggests the goal toward bilingualism rather than a priority to learn English, which can often come at the expense of the first language.

Ethnomathematics The anthropology of mathematics. Studies mathematical practice in all its forms, from academic mathematicians to mathematics embedded in cultural practice. Readily applied to mathematics teaching.

Fallibilism A branch of philosophies of mathematics holding that all mathematical knowledge is verified as true only because it has not yet been proven false. Contradicts absolutist philosophies of mathematics.

Feminism A movement spanning more than 100 years that aimed first to establish equality for women through suffrage (first wave), equality in home and workplace (second wave), and finally the undoing of gender norms as well as highlighting intersections to race, class, and other social identities (third wave).

Hidden curriculum The unstated experiences in school leading to particular socialization outcomes. Regarding class relations, a hidden curriculum prepares learners with working-class backgrounds for working-class work and learners with higher-class backgrounds for higher-class occupations.

Hook The opener to a mathematical lesson in which a community of learners is established, learners' prior knowledge is activated, and learners' curiosity is piqued for the learning about to take place.

Human capital The primary policy directive for mathematics education at present, in which mathematics teaching are learning is in service to corporations by preparing workers with the intangible qualities needed for profit.

Intersectionality A tenet of contemporary social theory rejecting a single-axis focus on social identities in favor of the multiplicity of social relations determined by several sociopolitically determined social groups. Highlights the systemic nature to domination and exploitation occurring throughout western industrial culture.

Meritocracy A mythology within capitalist logics suggesting that individuals fit within the socioeconomic order as a result of their talents and hard work. On the contrary, a critical view suggests that governmental and other structures exist to reproduce social class hierarchies, in which individuals are not given a fair means by which to advance higher on the social class hierarchy.

National Council of Teachers of Mathematics (NCTM) The primary professional organization for mathematics teachers in the United States and community for mathematics education research. Responsible for reform mathematics teaching and the new-new math.

Neoliberalism The current dominating school of thought framing social policy, in which free-market logics are applied to all sectors of life, including public education. Frames mathematics education as in service to corporate profit.

Platonism A philosophy of mathematics emphasizing that mathematical knowledge unearths ideal forms; it corresponds to believing that mathematics is a fixed and indisputable truth.

Poststructuralism An advanced branch of social theory that moves beyond fixed, rigid explanations of social relations. It articulates power as fluid and the result of intersecting social identities and sociopolitical contexts, including and beyond race, class, and gender. Individual and community experience can be described, but no objective truth can be determined.

Pure mathematics A major field in mathematics that defines imaginary spaces bound by a set of axioms and works within the space to make further claims.

Race A social construction of social identity groups. Originating as a sociopolitical ordering that establishes relations, power dynamics, and hierarchies between groups of people. There is no biological basis to understanding race.

Reform mathematics teaching This pedagogy resonates with the social constructivist philosophy of mathematics. It encourages classrooms to model the work of mathematicians, where learners problem solve, discover, conjecture, reason, justify, prove, and communicate mathematical knowledge. Also emphasizes "mathematics for all." The National Council of Teachers of Mathematics (NCTM) and the math education research community promote these teaching methods.

Response to intervention (RtI) The current method for special education that identifies learners with special needs and accommodates accordingly. This is critiqued by disability studies for its labeling of students with disabilities as fundamentally different.

Social class The social construction of class is the reproduction of a hierarchy of economic groups and general lack of social mobility. Social class levels are marked by one's own wealth, level of authority versus obedience at work, and amount of autonomy at work.

Social construct A concept held in society that acts as an indisputable truth yet in actuality has been constructed by communities and social groups. Typically used to refer to social identity groups such as race, class, and gender.

Social constructivist philosophy of mathematics Viewing mathematics as a social construct itself, this philosophy promotes mathematical knowledge as the product of practice by a community of mathematical knowers. Contradicts absolutist philosophies of mathematics.

Standard algorithm A routine process by which a school math problem is typically solved. Two-digit column addition and solving linear equations are examples. Mathematics education, typically conceived, focuses on the teaching of standard algorithms.

Task The central feature to an experience-based mathematics lesson. Reform mathematics teaching pushes us toward selecting tasks with higher cognitive demand.

Think-aloud An essential feature in which a teacher models mathematical skills, from algorithms to mathematical reasoning. In a think-aloud, learners actively engage when a teacher articulates out loud their thought process in problem solving, including all the information they receive, all the choices they have, which choice they make, and why.

Think-pair-shares An essential feature for active, student-centered mathematics lessons. A discussion prompt or problem is presented and each stage is given a time limit. Learners first engage with the prompt silently and on their own, next work in pairs to discuss, and finally are called on to share their findings as a pair.

Traditional mathematics teaching This pedagogy embraces an absolutist philosophy of mathematics and emphasizes rote learning, or memorization of number facts. The assumption is that mathematical skills should be taught without the conceptual understanding, which is reserved for being taught only once the skills have been mastered.

Translanguaging A critical viewpoint on language usage that rejects static, separated languages in favor of multilingual usages that maximize the efforts in communicating.

Universal design for learning Drawing from disability studies in education, a lesson and unit planning methodology that is inclusive for all students and disavows the labeling of learner needs and subsequent assumptions of inferiority.

White institutional space Mathematics education is deemed a white institutional space because the groups and individuals involved in its historical development are white and have acted largely in their own interest. Attempts by the mathematics education community to address racial injustice have been superficial at best.

White supremacy The sociopolitical ordering of relations between racial groups of people in which white dominance is asserted. False assumptions contained herein include innate differences in intelligence, athleticism, and sexual behaviors.

Whole-class discussion In an experience-based, reform mathematics lesson, an essential feature in which students share their findings during the task and negotiate meaning for mathematics.

INDEX